What people are saying about

Emergent

Our ancestors spread out of Africa in a series of waves and wherever we went we made a huge impact – usually driving the mega-fauna and their predators to extinction, and so, indirectly, transforming the landscape. But then, typically, the new ecosystems settled into new equilibria in which people and wild creatures lived in harmony – until the next disruption. Modern people are continuing the pattern, and right now we are in the negative phase but although recovery is still possible we no longer have the time or the leeway to leave it to nature. So what can be done and is likely to be done? And how might the world turn out, given that cause and effect in nature are decidedly non-linear, and ecosystems cannot be engineered precisely? Here is an original and seriously intelligent overview of the impact of humanity on the world at large – not just of Homo sapiens but of the entire genus *Homo. Excellent.*
Colin Tudge, Author and co-founder of the College for Real Farming and Food Culture

Emergent tells us what we long to hear – that we are nature. Miriam challenges, in a clear and concise way, the contemporary narrative of Human vs Nature, which has enabled us to dismantle the very ground upon which we stand. This book tracks the journey of that separation and reminds us of our true nature. It reminds us to tend our gardens as all living creatures do, as an integrated part of the beautifully complex and dynamic ecosystems we inhabit.
Caroline Aitken, Director, Teacher ar ˉ
Whitefield Permaculture and Co-Author
Garden

T0158662

This book brings new perspective in how to think about our environment and how we interact with it. For me, new to the farming world it makes sense of much of what I see and have experienced over the last 6 years starting up our small holding and trying to build resilience into what was very tired land. What's exciting though is that it also chimes beautifully with my world of strengthening and building wellbeing in the NHS. Understanding the holistic approach, that we must build and strengthen what is just another ecosystem in the NHS, is fundamental to positive change. If you need good soil to grow then the same applies to any ecosystem.

Professor Debbie Cohen, Emeritus Professor of Occupational Medicine School of Medicine Cardiff University

Emergent

Rewilding Nature, Regenerating Food
and Healing the World by Restoring
Connection Between People and the Wild

Emergent

Rewilding Nature, Regenerating Food
and Healing the World by Restoring
Connection Between People and the Wild

Miriam Kate McDonald

EARTH
BOOKS

Winchester, UK
Washington, USA

JOHN HUNT PUBLISHING

First published by Earth Books, 2022
Earth Books is an imprint of John Hunt Publishing Ltd., No. 3 East St., Alresford,
Hampshire SO24 9EE, UK
office@jhpbooks.com
www.johnhuntpublishing.com

For distributor details and how to order please visit the 'Ordering' section on our website.

ISBN: 978 1 78535 372 7
978 1 78535 373 4 (ebook)
Library of Congress Control Number: 2021945128

Design: Stuart Davies

UK: Printed and bound by CPI Group (UK) Ltd, Croydon, CR0 4YY
Printed in North America by CPI GPS partners

We operate a distinctive and ethical publishing philosophy in
all areas of our business, from our global network of authors to
production and worldwide distribution.

Contents

In the hope that we might all one day know ourselves as a strand braided into the fabric of the whole.

Foreword

There are many paths that may have led you to this book. What binds us all is unease, a sense that something isn't quite right. We are blundering into what can often seem a multitude of crises. Our climate is unstable and getting worse. The wildlife that we share this world with is disappearing at a rate that is unavoidably witnessed whenever we open our front doors. Alongside this we have the pervasive issue of poor mental and physical health across our nations.

Growing up within these crises I have felt uneasy, not only uneasy about what can feel like impending doom, but also uneasy about how we are dealing with these issues. We seem to have a penchant for superficial solutions to complex problems. The common way of talking about this is by likening it to sticking plasters on our wounds. If you were to look at the crises highlighted above, we seem to find ourselves in hospital with life threatening injuries and degenerative diseases. At hand are a suite of doctors, refusing to speak to one another or acknowledge their interdependence, each armed only with a box of plasters. One can't help but be uneasy about how many plasters are left in the box and how long they will stem the tide.

I set out on a bumbling journey to explore my unease. It is far too easy to point out the flaws in something; imagining alternatives is the hard part. Initially I made my way through the worlds of conservation, community development and agriculture searching for alternatives to the degenerative systems that we have put in place. I eventually stumbled into academia and in particular earth systems science. I became fascinated by the Gaia Hypothesis, the idea that the planet can be understood as a single system that is able to self regulate, much like our body regulates its temperature. This idea made total sense to me, throughout my practical and academic explorations I found a movement towards stability in nature. If we are consciously aware that Gaia exists and that we are playing havoc within it, what is our next conscious move? What is the role of humans within Gaia?

It was at this point in my journey that I bumped into Miriam, on a small farm near to the border between England and Wales. Here was someone walking the same path as me, asking the same questions, but had left from a very different starting place and had travelled a very different route. Miriam was exploring the ecological role of humans in our landscape; past and present. She was doing so in a very practical manner, far from my theoretical approach born from Gaia.

Sat up on a hill, looking at the marches rolling away in front of us, we spoke with excitement. Catching each other up on the paths that we had taken and what we had learnt along the way. In that one conversation I made more progress in my thinking than I had made in the previous year. I needed more of this, and I was to get it. Miriam and I began a friendship based on sharing knowledge and a passion for understanding.

The reason I have been drawn to Miriam and her thinking, and I feel you will be too, is the way in which she validates our unease and explores why we have ended up in this place.

I feel this book is a distilling of this thinking that goes even further, to offer insights into where we may go from here and how we may act on our unease.

This book helps to highlight that what we see as separate crises are actually entangled with one another and at the heart of our climate, biodiversity and health crises is an identity crisis. For so long we have viewed ourselves as separate from the world around us, able to choose when and how we interact with it. It is this world view that allows the deconstruction of our planet, the deconstruction of ourselves and, in essence, the deconstruction of our whole.

Miriam emphasises that there is an alternative world view, one in which we see ourselves as participants in a complex web of life. This world view accepts that we are inextricably tied to the world, every one of our actions garners a reaction and in turn these reactions feed back into our actions. If we take this world view and walk forward with it we can begin to reconstruct, or restore. Through altering our sense of identity and finding wonder in being something small, part of something massive, we can start to restore our climate, wildlife and selves all at once. We can restore our whole.

Far from being a solely philosophical attempt to reconcile ourselves with the world, this book is a practical exploration of science, history, language and land management. It will join you on the path that brought you here and walk with you, introducing you to how and where we have always been participants in our landscapes, and it will look forward to the horizon with you.

Miriam, and this book, do not claim to have all of the answers; what they do is raise questions that I, and likely you, have never asked. It is a calling for others to ask their own questions, in a new way, and in doing so be part of restoring our whole.

I feel it is now time to stop and let you read the book and meet Miriam for yourself. I am writing this as a friend and colleague of Miriam's but more importantly as a fellow uneasy traveller on the road. I hope this book brings you a fresh perspective on your journey, as it did me, and that you can share these insights with others you meet along the way.

Robert Owen
Independent Researcher and Visiting Fellow at The University of Exeter and Co-Director of Holistic Restoration

Introduction

When I was young, someone told me a joke. I didn't understand it at the time and it played on my mind, turning over and over in my head as I tried to understand what it meant and why it was funny. I still remember it to this day, it goes like this;

Two young fish were swimming along and they passed an older fish swimming in the opposite direction. The older fish said, in passing, 'How's the water up ahead?' The young fish kept on swimming and, once the older fish was out of earshot, exchanged glances and said to one another, 'What's water?'

I now understand what this means, although I'm still not entirely sure why it classes as a joke. There are so many things in our everyday lives that are always there, and have always been there since we first drew breath that we fail to see them at all. The fish, growing up in water, lacked a concept of what it was. And people, growing up swaddled in belief systems, landscapes and languages, are no different. There is so much of the very foundation of our world that completely passes us by. We accept that we are people, that we have dogs as pets and cattle as livestock. That the birds we feed in the back garden are wild and the dandelions that sneak into our patios are weeds. We subconsciously categorise our world as our parents and grandparents did, often without even knowing we are doing it, and accept the 'water' in which we swim.

Beyond our acceptance of our shared water we also seem to fail to notice when it changes. Over recent years shifting baseline syndrome, or environmental generational amnesia as it is known in psychology, has been pointed at to explain people's extraordinary lack of response to creeping changes. Shifting baseline syndrome was first coined in 1995 by Daniel Pauly in relation to diminishing fish stocks. In general younger generations, growing up in an already degraded world with lower fish stocks, see the level of wildlife and fish available today as less shocking compared with those of the older generations who remember a time when both fish and wildlife were more abundant. With each generation that passes a new baseline, a new idea of normal water conditions as it were, is adopted and over successive generations big changes can occur with no one truly appreciating just how large they are, or even that they are occurring at all.

I have spent over a decade delving into our shared water, all of the things that we accept without even knowing that they exist, and a striking pattern has emerged. Over the centuries the creep, below the level of detection, has been leading in one direction; towards division.

In the first section of this book I am going to tell the story of a baseline that has shifted so slowly towards segregation as to not be recognized by many who have lived, and are living, through it. To take a look at the steady plod that has carved life up into ever smaller fragments and to delve into what some of the consequences of these divisions have been.

The second part of this book will focus on our present. We will explore where and how the damage done through division is being repaired. We will take a whistle stop tour of a new science of complexity that has, at its heart, an acknowledgment of just how complex and interdependent life is. This interdependence has guided much of ecology over the years and we now have some amazing insights into how our landscapes function together with life to draw from their integration a rich, dynamic and diverse tapestry that ripples across our world. From here we will discuss what strides conservation has made, what we can learn from rewilding and whether there is a place for us within such a vision of the wild. We will discuss our ecological roles in ecosystems and assess whether we can contribute positively, taking up a beneficial role in ecosystems once more. From here we will delve into regenerative agriculture and learn how farmers are reweaving elements that have long since been separated to alleviate floods and droughts, build soils, heal relationships and feed us now and into the future.

The final part of this book will focus on where we can go from here. Whether it's possible to reintegrate people fully into the wild and the wild fully into our lives and what this might bring with it. At every stage of reintegration unexpected things emerge. Growing beans with wheat or grass with trees yields far more than an addition of grass to trees or wheat to beans. Emergent properties spring from integration, from complexity, and as we foster increased connection we should expect the unpredictable. We will close by imagining a reintegration of the final and deepest division in our world, the divide between human will and the will of the wild. And this, the oldest and first division, is where we will also begin our book.

Part I

Our Divided Past

Chapter 1

The Changing Face of Britain

Standing on the windswept shores of Skye at low tide the footprint was barely discernible but there it was. A dinosaur family had left indentations in the soft sand of its home which had been pressed and preserved into the rock to be revealed 165 million years later. The British Isles are old and have not always assumed the shape or position on the Earth that they occupy today. Around 500 million years ago England and Wales were below sea level and attached to what is now Scandinavia but Scotland was attached to North America, and an ocean separated the two. Gradually the two plates collided and where they met mountains were created that would later go on to become the Scottish Highlands, including Ben Nevis, and the Scandinavian mountains.

The fusing of the plates caused the land mass to push out of the water emerging into the air of the southern hemisphere at the latitudes where deserts form. The new mountains started their long history of erosion and the desert winds blew deposits into areas that would form the Old Red Sandstones still visible today. Over this period the super continent of Pangaea was still forming with Northeast Africa pushing into Southern Europe causing mountains to form along the collision. The south of Britain lies over these mountains of igneous rock that have been weathered and exposed as the hills of Bodmin and Dartmoor. The same upheaval caused the mineralised deposits of copper and tin to form across Devon and Cornwall that would go on to support so much of their mining industry.

These collisions moved the British Isles out of the southern hemisphere towards the equator. Initially under a productive and warm sea vast quantities of marine creatures deposited

9

their shells to form the limestone found over the Mendip Hills, Wales, the Peak District, along the Pennines and into southern Scotland.

Eventually Britain rose from its watery repose ascending into an equatorial swamp forest of vast tree ferns inhabited by giant dragon flies, millipedes, scorpions and spiders. Forests covered the land growing profusely in a time before any advanced herbivores had evolved to browse them. Copious quantities of leaves, branches and trunks rained down over the years and were crushed into our coal deposits.

Britain continued its journey north leaving its bountiful equatorial position carrying its precious cargo of would be coal with it. As it passed through the desert latitudes of the northern hemisphere it was once again sand blasted before moving towards a more temperate position. By this time the early reptiles that crept out under the cover of the swamp forest and had evolved, and evolved again, to mature into dinosaurs. The footprints of the Skye dinosaurs which my hands had traced were left in an almost unimaginably different world to ours when Britain was a part of the super continent, Pangaea, before it started its long process of disassembly, tearing open to let the Atlantic gush in. Sea levels rose and fell over the ages washing marine life back and forth and as the continents approached their modern day resting places Africa jostled into Europe crumpling the earth's surface into something like its modern day arrangement of mountains.

Britain was now in a fairly northerly position exposed to a fluctuating climate that propelled glaciers back and forth across the landscape for 2.5 million years eroding mountains and forming great rivers of melt water that transported huge quantities of fresh sediment across the landscape forming deposits of sand, silt and mud. The regular grinding of rock by glaciers and redistribution of sediment by rivers allowed deep soils to accumulate, rich in all the minerals needed for life.

As glaciers swept to and fro vegetation and animals were buffeted back and forth in their wake. As the ice advanced more and more water became locked up in glaciers and less and less was available in liquid form to fill the depression between Britain and Europe. The sea bed of the English Channel, known as Doggerland, was exposed and plants and animals were free to migrate as they pleased across it. Giant Irish elk (like supersized red deer), woolly rhinos, muskox and woolly mammoths alternated with horses, bison, straight tusked elephants and narrow-nosed rhinos. Hunting these huge grazers and browsers were scimitar toothed cats, cave hyenas and the giant cave lions.

Standing in Creswell Crags on the Nottinghamshire Derbyshire boarder it is almost possible to see how these lands may have looked tens of thousands of years ago. Creswell Crags is a limestone gorge with networks of caves on either side of it hewn from the rock as a glacier to the north melted sending water cascading downwards to form a mighty river. Trees now grasp the banks and the stream has been blocked to form a lake on the valley floor but it's easy to image the hulks of rock, bare in a glaciated world, with short grasses trickling between them. These caves have always provided a secure home with hyenas lurking here, leaving to prowl the ice sheets to the north in search of woolly rhino, mammoth or reindeer carcases during ice ages or horse, hippo or elephant remains in interglacials. Humans have also used these caves, Neanderthals lived and hunted down the valley 40,000 years ago and *Homo sapiens* dwelt here 22,000 years ago etching images of bison, reindeer, birds and horses onto the walls and engraving them in bone fragments.

People, or more accurately different species of the genus *Homo*, have become embroiled, just like everything else, in the glacier dance. Nearly 1 million years ago some ancient cousin of ours used stone tools on what is now our island and 500,000 years ago *Homo heideilbergensis* set up camp here more meaningfully. Over the years that followed Neanderthals wandered with the ice ages in and out of Britain at least 8 times, possibly following the herds of migratory megafauna, as they shifted their ranges in response to the ever moving frozen tide.

Homo sapiens spread out of Africa about 100,000 years ago and radiated out across Eurasia following in our cousin's footsteps. Unlike previous species of *Homo,* however, wherever we went, others vanished. On the continent and across the rest of Europe all of the other species of human, as well as many of the Eurasian megafauna including the straight tusked elephant and narrow nosed rhino, the mammoth and giant elk, vanished. Modern humans did not stop their dispersal at Eurasia but crossed into Australia about 47,000 years ago and clambered into the Americas possibly around 14,000 years ago (although the timing of this is hotly debated). In these regions too the first human occupation is linked to massive declines in the native megafauna as the vast marsupials of Australia and the giant ground sloths, camels and horses of the Americas vanished from the landscape. It is possible that these new and effective hunters targeted the largest lumps of meat first which, having never encountered a tool using ape before, did not know to avoid us. This idea is known as the overkill hypothesis and is highly regarded, although still disputed with the main alternative being that the extinctions were linked to changes in climate. Whatever the cause, across the world, as people entered ecosystems many other species vanished from them, profoundly altering how those ecosystems hung together and what sort of landscapes existed.

It takes a vast swath of time to highlight this first division. As people expanded across the globe we most likely had a hand in dividing the world from its megafauna. This first act was likely the result of one of the very first shifting baselines. David

Nogues-Bravo, whilst trying to untangle the climate effects from human hunting effects in mammoth extinction, estimated that a person would have only had to kill one mammoth every three years to result in their eventual extinction, and that estimate comes from assuming mammoths were fairly abundant in what remained of their ice age world. If that generous assumption is wrong a person may have had to take only one mammoth every 200 years to result in their extinction! It is likely that the people living through these times were not aware at all of how the water they were swimming in was changing.

At the end of the last glaciation everything in Britain was ice above a line from north Wales to the Wash with a treeless arctic grassland covering the lower part of the country. Over the last 15,000 years the climate warmed and a swath of birch and pine forests marched north, closely followed by hazel and aspen. In this interglacial, however, the large megafauna of elephants, rhinos and lions did not follow having been driven to extinction or near extinction on the continent in the meantime. Our trees, faithfully returning, came back to a very different landscape. One in which they were browsed far less fiercely.

Gradually as the climate warmed and pioneer trees softened the landscape oak, lime and elm arrived. By 6000 years ago the British Isles was possibly clothed in the wildwood which may have grown fiercely, taking advantage of the sudden lack of large herbivores. Pine and birch possibly dominated much of Scotland and oak and hazel much of England extending up into lowland Scotland and Wales. The warmer and drier east and south of England gave way to extensive lime woods that pushed north west into the midlands and hazel and elm woods dominated warm and wet Cornwall and south west Wales. Interspersed amongst the trees were our high moorlands with their thin, acidic soils often perched atop impervious granite, strung with a network of oak and hazel woodland and heather and peat bogs.

A tangle of rivers, lakes and wetlands wove through the rich clay vales and spilled between willow and alder carr housing diverse assemblages of plants and invertebrates and leading out into wet grasslands still browsed, grazed and coppiced by our remaining, although much smaller, herbivores; elk, deer, beavers and aurochs (the wild ancestor of cattle) among them. Oliver Rackham estimates that over 25% of the British Isles would have been covered by wetland of some form or another from the acidic upland peat bogs to the highly diverse, alkaline lowland fens and probably even more would have been seasonally inundated as sinuous rivers freely spilled their banks. Where tussocky vegetation and scrub met open water high levels of fish and invertebrates could bloom supporting huge numbers of birds including pelicans, storks, cranes and night herons.

The shifting river homes of beavers opened up networks of wetland glades and ponds that created highly complex habitats hosting a wealth of other species. As beaver ponds silted up and their residents moved on to streams new, fertile grasslands would have sprung up providing grazers with a veritable feast of succulent growth. Bears too would have gorged on the tubers, fish and berries that would have sprung up in pools and woodland clearings, transporting and planting fruit tree seeds in their dung as they wandered the landscape.

Through this wilderness hunted our remaining predators, wolves and lynx, as well as the more familiar fox.

This was already a changed land, however, the loss of the elephants and the rhino, the large carnivores and the Neanderthals, had impacted our ecosystems and how they functioned. The arrival of a very keen little ape had reshuffled our food webs and landscapes creating new rifts within old ecosystems.

As a species we, people, have never known a complete world. Even before we had begun daubing colour on the walls of caves we had radically altered our earth. Severing the ties between our landscape and the giant animals that had sculpted it through their browsing and hunting. This is an incredibly important point, and one that does not get the air time that it deserves. Literally everything that we have ever known or recorded on earth, everything that science has ever uncovered about our biosphere, has been discovered in a disturbed world. This is the first thing that we need to bear in mind when thinking about the water in which we swim. It is turbulent, we have grown within the flow of the water and our lives have been created by, and have contributed to, its constant movement. Movement that was initiated tens of thousands of years ago.

This first division affected our world fundamentally but soon we would take these interactions up a notch, to not only replicate some of what these ancient species had been doing but to sculpt our own species to do it in their stead and to begin the slow process of pushing what was probably an unusually wooded landscape back towards dynamism and then further into a highly disturbed and divided future.

Roots of Revolution

People have a deep, biological, connection to soils. The smell of damp earth, or more accurately the molecules released by soil bacteria living in damp earth, stimulate a release of serotonin in the brain that lowers anxiety and depression and promotes mental wellbeing. The human nose, and many non human noses for that matter, are incredibly sensitive to these chemicals including one called geosmin which can be detected at 100 parts per trillion and has been linked to our ability to tell which habitats would have sustained us. Damp soil is life.

All around the world members of the *Homo* genus would have dug through soils to find tubers and would have left bare patches of earth into which annual seeds would have fallen. As *Homo sapiens* replaced other members of the *Homo* genus and the megafauna across most of the globe we too would have engaged in digging for tubers, collecting seeds, nuts and fruits and hunting or scavenging for meat.

In the deep past our ancestor species had discovered fire and we used this to open the door to a new world. We are very familiar today with the central role that fire played in our evolution, allowing us to cook food to liberate more calories from it, keep us warm and modify our landscapes. But this familiarity masks a massive underlying change. Fire meant that we were no longer passive recipients in the world but active creators of it. In order to use fire to the best of its potential we would have needed a concept of the future and also the ability to recognize that the future could be affected by decisions in the present. Fire was a tool that could be used to burn areas of old vegetation to attract game *in the future*. Wild seeds could also be intentionally gathered, planted and tended to increase fruits, nuts and seeds *in the future*. The truly colossal shift from being a passive recipient of the world in the present to being conscious of the potential possibilities of the future opened the door for agriculture.

Many modern hunter gatherers will burn landscapes to maintain good grazing for game and will also plant and tend native, edible plants creating managed landscapes across vast areas. The foragers of the Kalahari will go out on foraging trips gathering tubers, roots, seeds or fruits to bring back to temporary camps. After a few days, when the camp is left, any uneaten food that is either past its best or cannot be transported is deposited as waste in middens or else is planted and left to grow for the future. On future foraging trips, old camps are revisited and over the generations networks of these food filled oases build up across the landscape. It is not difficult to see how agriculture could spring from such oases, as tended wild food sources become domesticated crops or game attracted to intentionally burned lands captured in fields as livestock. Indeed when researches look for the first signs of domestication of a species of livestock one of the things that they are looking for is decreased adult size. This is also what happens, however, within regularly hunted wild populations, as people eat the larger individuals the species as a whole tends to get smaller over time as only the smaller individuals get chance to breed.

There is no hard and fast physical line between hunter gatherers tending wild foods and farmers tending domesticated ones.

As soon as people have a concept of the future, it would seem, we start to make decisions in the present that we believe will lead us towards a bountiful time to come. Although the practises of tending wild foods and rearing domesticated ones are very similar, at least initially, the *beliefs* underpinning them seem to alter considerably. The difference between foraging and farming therefore seems, to many including myself, to rest on how people view themselves within the world far more than on what people are actually doing practically.

Maybe 70,000 to 100, 000 years ago, well before the dawn of agriculture, the human brain evolved into something that is quite possibly unique within the world. What exactly happened and why is still relatively mysterious despite the best efforts of many scientists around the world. What we do know, however, is that our brains seem to have evolved in a way that allowed us to see meaning in the world. We could grasp the idea of ourselves existing as individuals and also understand that others exist too, with their own agendas, ideas and understandings. We started to try to communicate our interpretation of the world to others and to read into what others meant by what they said and did. We also looked beyond our human companions, however, to find meaning in other animals, plants, the wind and the rocks. We started to live in a world of meanings, and language allowed those meanings to be shared among the group and passed down through the generations resulting in the development of 'culture'. Culture is learned by each person alongside their language and forms a major part of the water in which we swim, sculpting how we see and interact with the world. We think in the words given to us by our language and we see the world using the metaphors given to us by our culture. It was this composite that shifted out of all recognition as we moved from hunting and gathering to farming.

Hunter gatherers tend to have languages and belief systems that see and describe the world as one and alive. People play their part alongside the rest of life, including animals, plants, rocks and sky, for example to sustain their world. Hunter gatherer belief systems tend not to distinguish between the 'individual' and the 'group' or between 'people' and the 'wild' basing their worldview on the understanding that an individual cannot survive beyond the group or people survive outside of the wild. Hunter gatherer belief systems rest on the assumption that the world is an ever giving parent, continually gifting life.

As people settled into agricultural life, however, the language and belief systems shifted, gradually weaving together another culture.

While time, planning and decision making are used by hunter gatherers there is less emphasis placed upon them possibly due to the greater level of trust that hunter gathers tend to have in a generous and maternal idea of the world. In agrarian societies, however, planning and time both have to be engaged with very heavily to ensure that sowing and harvesting are not missed or mistimed. Time itself distils out into a past, a present and a future which can lead off in several directions dependent upon actions taken in the present. In agrarian societies there is a swelling of an idea of self as an individual and the importance of individual actions taken in the present. The agrarian view of the natural world shifted gradually from one of eternal parental giver to one of a wild and unpredictable 'other' who's wits the farmer was pitted against. Nature became the threat, the unpredictability that could bring pests or natural disasters. It turned from bountiful and eternal giver to scarce and eternal threatener. The idea of the self was teased from that of the group and the idea of the civilised was extracted from that of the wild.

As agricultural societies aged, the idea of the self matured. Settled agricultural areas could, for the first time, amass food as wealth and it's probable that the more successful individuals rose in status allowing stable hierarchies to develop and warrior classes to emerge to protect and enforce those hierarchies. Once wealth had been amassed the anxiety of losing it rose ever higher and the desire to subjugate the unpredictable wild must have bloomed in response. The individual's *will* emerged and it was set against the will of the rest. Farming was to sculpt that will into a force to be reckoned with; civilisations and individuals that fought tooth and nail clawed their way to the top guided by their desire, their will. Farming opened the door to the belief in the great divide between humanity and the rest of the universe and fostered the idea that it was the individual and people who could secure a future against the savagery of the wild, rather than the wild gifting continued life. Farming then gave birth to the second great divide, the self fulfilling prophecy that people are separate from nature.

An Island Home

Although it is a blurred line farmers gradually replaced hunter gatherers. Farming was born in over 14 different places around the world, each with its unique set of domesticated plant and animal species selected from a very wide array of wild habitats and species. The form of agriculture that would one day reach Britain, and eventually the rest of the world, originated in the Fertile Crescent. It had been born in the soft alluvial soils of river valleys that had provided ample water, renewed fertility with sediments and periodically drowned both pests and weeds. It has been suggested that it was the ancestors of broad beans that were gathered and poked into the soft, rich ground of this first nursery over 10,000 years ago eventually leading to the domestication of wheat accompanied by cattle to till and prepare the fields in which it would one day grow.

As agriculture spread from its river valley nursery across Europe it had to adapt and as it arrived in Britain, Britain had to adapt to it. Water was readily supplied by our gentle rains but fertility was not as easily maintained nor were weeds or pests suppressed away from the sediments deposited by the great rivers of the Fertile Crescent. The first farms tended to migrate across the landscape with their new crops of wheat, barley, sheep, goats, pigs and cattle, clearing and tilling areas to mimic the soft and open river soils of the Fertile Crescent in which to grow cereal seeds. Unfortunately without the renewing flow of sediments from upstream these areas lost fertility over time and the fields were abandoned to be reclaimed by nature and the farm moved on.

The new cattle and pigs that arrived in Britain slotted into our mottled landscapes of woods and clearings seamlessly and this is not to be marvelled at because their wild cousins, although of slightly different bloodlines, had been creating those clearings for centuries. Aurochs and boar are both native to the British Isles but were domesticated many hundreds of miles away, in the Fertile Crescent in the case of aurochs and in Turkey for boar. The new pigs interbred with our native wild boar, creating a distinct band of breeds that have survived to this day, and the cattle precipitated a new way of life. The arrival of cattle probably encouraged people to switch from a life following the seasonal migrations of their prey to a life of pastoral nomadism as they herded their new charges down old routes. Both species were well suited to the food and landscape on offer and bred well, especially pigs, multiplying through our gentle landscapes and aiding their human collaborators in opening up new grazing and feeding areas allowing farms to spread outwards.

Standing in the cool air of Ennerdale or the mottled landscape of Knepp one can almost feel the shadow of the monstrous and now extinct aurochs looming behind the cattle as they browse and it is not too hard to imagine the impact that their new, diminutive relatives would have had on our lands when allied with their human partners.

This small scale, gentle ebb and flow of human interaction probably sculpted a landscape not dissimilar to that created in previous interglacials by megafauna. People cleared trees much as elephants would have done, cattle grazed and pigs rootled.

Gradually the impacts of human clearing and pig and cattle foraging opened up more and more pastures to be grazed by multiplying stock. Over 3500 years ago permanent clearings were established with the first true field systems and heath lands allowing arable area to increase and sheep to be shepherded in the more open places but also meaning that farms could no longer easily migrate to fresh, fertile soils. New methods of maintaining the productivity of the land were needed and with the introduction of the scythe in Roman times it became possible to cut and store grass for winter feeding indoors. Stock could be housed over winter accumulating a rich mix of manure and old hay beneath them that could then be spread on the land over spring raising the fertility of fields used for cereals or forage production.

An intricate mix of wetlands, coppice, wood pasture, small arable fields, meadows and pastures spread across the country and for thousands of years these patterns of land use changed very little and only very slowly. This was a land driven by human will but slowly, sleepily. It was compelled by no big ideas or great desires and the lives of peasant farmers came and went, altered only marginally by the wars of the elite washing over them.

The Enclosure of Food

The Norman conquest of Britain in 1066 shook up our landscapes once more and disrupted the steady plod that life had settled into. When William the Conqueror won the crown from Harold II, William rewarded his loyal noblemen with vast land holdings, much of which was in the uplands. He also strategically took control of the church and the lands that came with it and awarded further areas of the uplands to the monasteries.

Wool had been a highly prized commodity since Roman times and the introduction of a new breed of sheep that had spawned both our short and long wool breeds had only fuelled this desire. By the time of the Norman Conquest British wool breeds had been crossed with much older, tougher breeds of hill sheep to produce a hardy, upland fleece producer. The new Norman land holders quickly realized that their lands were well suited to raising these sheep for their precious covering of wool which could be exported for good prices to the continent. Over the 1200s to 1400s the wool trade boomed and sheep multiplied rapidly through our uplands, often at the expense of the people who had lived there previously. Local villagers had traditionally herded their cattle and sheep up onto the hills and valley heads for the summer months, a practise called sheiling in northern England and Scotland, where they grazed and produced milk that was turned into cheese for the winter. The displaced people, no longer able to utilize the resources of the uplands, went in search of sustenance to the towns that were springing up across England to process and export fleece to the continent. Large numbers of workers amassed in the budding cities, all of whom needed feeding and most of whom had their new wages at the ready with which to buy food. For the first time landless people worked for a wage with which they bought back the sustenance that they had lost.

Upland land ownership and fleece production set in motion a division of people from the land that would trickle down from the hills to flood across the lowlands, segregating people from sustenance across Britain. This division was seen, not just as a severance of the people from their form of sustenance but from their identity, from their very selves. The land was a part of them and they were a part of the land.

The Enclosure Acts, which were the formalisation of this across the lowlands, bound the landless poor to the role of cheap labourer in the cities and freed the land owning elite to experiment in squeezing the most out of their estates.

William the Conqueror precipitated a new way of viewing land and he opened the door to the exacerbation of the concept of the self, the possibilities available in the future and the degree of control that could be exerted over both people and nature. Land was increasingly seen, not as a shared resource to be collaborated with and from which sustenance arose, but as a commodity to be used and from which money could be extracted for future personal gain. Life was separated from land with money being used to bridge the ever widening gap in between. At the time it is very unlikely that people could have foreseen just how long lasting and just how calamitous this rift between land and life would prove to be.

A Divided Science

Science is very old and has grown through many incarnations to become the trusted backbone of our society today. It is, however, just like everything else, only an idea and is as subject to subtle sculpting forces as anything else. Jeremy Lent in his fantastic book 'The Patterning Instinct', has explored the evolution of science in great detail and what follows merely scratches the surface of this fascinating story.

Divinity, in ancient Greece, was generally believed to exist in the heavens and only through logical reasoning could humans approach it. Philosophers debated the biggest questions of the universe in almost gladiatorial battles of logic with the one able to demonstrate the most solid reasoning being crowned the victor and leading, over time, to the accumulation of logical inferences about the divine 'truth' of the world. The emphasis placed on reason being required for divinity meant that the human bodily experience, as well as the rest of the natural world, could not be divine because it lacked human reasoning.

Christianity inherited this divided world view but it also wove a new strand into an old tapestry; the idea that we fell from Eden; a place in which we had command of the animals of the earth and the birds of the air, and that it was our birthright to regain our place of mastery.

Science grew from these root assumptions, that the world lacked divinity, that the truth can only be found through logical reasoning and that people are objective observers of a separate creation over which we can have dominance.

Influential French born philosopher and mathematician Rene Descartes expanded on these ideas when he wrote in 1644:

it is as natural for a clock, composed of wheels of a certain kind, to indicate the hours, as for a tree, grown from a certain kind of seed, to produce the corresponding fruit.

Descartes' world was a mechanical one. He believed that the universe was like a giant clock and worked upon simple, linear rules of cause and effect. One cog moved another cog which was destined to move a third and if enough of the cogs could be uncovered the workings of the world would be unveiled. He sought the mathematical equations and fundamental laws upon which he believed the earth to run, even disassembling live dogs to find out how they worked and comparing their whines to poorly oiled cogs in a machine.

The metaphor of the world as a clock was a powerful one and one that led scientists to attempt to take systems apart to see how they worked and rebuild them anew to meet our own designs. It was built on a foundation of questioning and had a new positive, prospecting attitude which opened the door to radical reform. People set out across the world to discover new ideas, places and riches and money started to be lent, not based on someone's track record for paying it back, but based on what someone might find in the future. A world full of potential opened up before us and science, allied with capitalism, was perfectly placed to take it. The new faith rested on the hope of future riches to pay back debt in the present and science took amazing strides forward towards a bright new world full of possibility.

Reductionist science has been enormously successful, bringing us massive advances in every field resulting in tremendous impacts on our daily lives but this has come at a price. The divide between humanity and nature has been deepened and the power relation skewed and masked.

Brave Industrial World

It wasn't long before the desires of landlords to reap the financial benefits of land ownership merged with the new and scientific view of the natural world as clockwork machine. Together these two ideas birthed a new form of highly productive agriculture over the 1700s and 1800s feeding a rapidly inflating population. New varieties of wheat and barley were bred and sophisticated machines built to harvest and thresh more grain. New and powerful horse breeds were also harnessed to modern machinery to drain wetlands, reseed uplands and convert heaths into arable fields rapidly increasing the area of land that could be worked intensively.

To achieve these gains land owners started to think differently about their estates. They began to identify a certain, far more restricted, set of enterprises that had the potential of being most profitable on their ground and tailored their operation to produce them optimally. The idea of singling out 'the best part' of what had previously been a system of food production grew in prominence as traditional mixed farms of livestock, pasture and arable were replaced by modern units specialising in just one or two enterprises.

The traditional mixed farms had maintained fertility by accumulating manure over winter from housed stock to be spread on arable fields and by rotating tilled fields with grazed ones, giving the land some time off to recover between ploughing. As arable operations were split from livestock enterprises these beneficial relationships were severed and the old issues of maintaining fertility and suppressing weeds away from a sediment depositing river started to rear their heads again.

War and the Enemy

The First World War marked the next major upheaval of British Agriculture and came at a time when Britain was importing around 60% of its food, including meat, eggs and cereals, from around the world. A huge exodus of both people and horses impacted the country leaving farmers with limited power to produce the vast quantity of food required at a time when our imports were being cut off. Arable production had been decoupled from livestock grazing or fallowing in certain areas for many years by this point and soil fertility had declined rapidly. The answer to the age old problem was sought in a new source of understanding, science. A German chemist, Justus von Liebig, had proposed a chemical basis to plant growth over the 1800s, he believed that plants required specific nutrients to varying degrees and the nutrients in lowest supply would limit growth. He called this the *'law of the minimum'*. By careful application of the limiting factor, fertilizers, he could enhance growth and in trials his fertilizers did magnificently well.

Fertilizer factories were built and started pumping out chemical fertilizers, however, British farmers were largely sceptical of the new methods of farming and uptake was slow with many fertiliser plants struggling to make them pay. This all changed with the War, however, as the huge strain on food production was felt and farmers scrabbled to raise yields from depleted soils. The First World War drove home the potential of both tractors and chemical agriculture to boost yields and these trends were further driven home by the Second World War. Modernisation was seen as key to supplying the yields required and vast quantities of land were ploughed up and fertilised for cereal production. Tractors continued to increase in number as horses declined and many farmers were aided in acquiring a tractor and modern farm machinery to go with it by the government, allowing them to increase yields but also fundamentally altering how their farms functioned and tying them into a new form of agribusiness; if you had invested in a tractor or other machinery you needed to make it earn its keep by up scaling whatever it could do most profitably.

Between the World Wars and after the Second World War there was a very obvious need for agriculture to yield highly, to underpin the rebuilding of Britain, but there was also another need; to encourage farmers to keep buying fertiliser. The factories that created the nitrogen based fertilisers for agriculture were the very same ones that created the nitrogen based munitions of war and if war was to descend upon Britain again those plants would be needed, straight away. Farmers were required to bankroll fertiliser manufacture in the meantime.

As the Second World War came to a close farming settled down once more but into a completely different shape, one focused on enhanced yields and simplified, mechanised systems. Not only had people and horses been largely stripped from the land, replaced by tractor powered machinery, but so too had all life that wasn't the crop or stock of choice. There was no room for an untidy field margin or an area of boggy ground; every inch had to pay its keep and maximum productivity became a matter of national pride. We asked for a simplified and highly productive food system and farmers delivered spectacularly.

The nation was also gripped by a strong idea of 'the enemy'. The enemy of life could be the Nazis but it could also very well be the weeds stealing yields from hungry mouths, pests guzzling down crops or the rest of the natural world, so famously by now 'red in tooth and claw'. Not only were the less profitable parts of what had previously been a system ignored, they were now actively persecuted. Complex and balanced systems were stripped back to individuals and the individuals carved up into good and bad. A combative approach to everything that was not a crop had been born and the divide between human and nature refined once more.

Industrial Cereals

Cereal production exemplifies perhaps better than any other the immense changes that have occurred in food production over the last couple of hundred years. The cereals that we have relied upon in Britain for most of our history have been oats and rye, tough plants that could yield in less than optimum soils. When agriculture first arrived in Britain whole farms moved across the land, leaving behind them weeds and pests and moving into fertility created over thousands of years by wild processes. As the land filled up farms ground to a halt but fields could still move, a practise that became known as rotation. Soil fertility in one field is consumed by growing a cereal crop whilst other fields are rested and grazed by livestock to raise it once more. Over the 1800s the rest periods diminished in length to allow more land to be in the arable part of the rotation. The cereals themselves also started to change with local cultivars gradually being replaced by higher yielding equivalents and rye and oats swapped for wheat and barley. Both types of cereal were fertilized more heavily to replace the long rest periods that had been lost from the rotation and to stimulate greater yields resulting in the wheat, which had previously been adapted to lower nutrient soils, growing tall and fleshy with a heavy head of grain predisposing it to falling over. After WW2 the first of the high yield varieties was bred with a short stem that could withstand incredibly high applications of synthetic fertilizers without falling over. Its reduced stature resulted in it shading the ground less thoroughly beneath it allowing weeds to grow up and compete with it which in turn necessitated applications of herbicides. The large areas of lush and fleshy wheat monoculture lead in turn to the potential for pest outbreaks and the need for large scale applications of pesticides, fungicides and other treatments to ensure the new crop was protected.

By teasing apart the web of connections that had held cereal production within the farm system, and exposing the now naked crop to the inspection of a global market, yields sky rocketed but so too did unforeseen problems. The trouble with dividing up complex systems is that, once you have snipped out the bit that you want, you are left with a whole host of loose ends that can unravel in truly unexpected ways.

It took many decades for these threads to work their way back through ecosystems and their effects to be spotted first by Rachel Carson and then by Robert Van den Bosch. The application of pesticides and herbicides triggered a set of reactions from the ecosystem into which they were poured, termed the pesticide treadmill, which farmers now found themselves caught on. What were at first mild applications of pesticide killed off not only the pest species but also their natural predators which were far slower to rebuild their numbers than the pests and much less adept at migrating in from surrounding intact ecosystems. It also became apparent that what had previously been benign species in fully functioning ecosystems were all too happy to step into the role of pest once it had been vacated by its former occupants resulting not only in more pest individuals but also more species playing the pest role. The farmers then found themselves facing higher pest levels, freed from their former competitive and predatory restraints. The pests also started to evolve resistance to the repeated doses of pesticides prompting research efforts that produced deadlier versions. An arms race had begun between researchers and pests with both farmers and wildlife as bystanders caught in the cross fire.

The problems did not stop with pesticides, a similar vicious cycle developed with weeds as the herbicides applied wound their way up the food chain accumulating in higher trophic levels causing all sorts of unforeseen side effects. Fertilisers too broke down the natural mechanisms that released nutrients to the plants, killing off the mediating microbes and other soil life and allowing organic matter levels, the soil's natural nutrient bank, to slip lower and lower. The replacement of the delicate natural system of nutrient release by artificial fertilizers tied farmers into higher and higher applications of synthetic fertiliser as their soils became addicted to the inputs. Farmers also had to over apply fertilisers because, unlike the natural drip feed system that was highly responsive to plant growth rates, artificial fertilisers were applied in one big glut often poorly timed with plant growth. A lot of the fertiliser leached from the soil before the plants could make use of it spilling into our waterways and then washing out to sea poisoning life as it went.

The efforts of plant breeders and chemical manufacturers, often the same companies, resulted in vast quantities of cheap grain flooding the market, their forerunners leading partly to the Great Depression of the 30s, and rapidly producing a surplus of fertiliser and grain in the aftermath of the War. Even as cheap grain lowered prices around the world, driving people from farms into cities in search of work and governments struggled to implement subsidies to prop up their agricultural systems, research focused on how to increase yields still further and the modern story of 'we need to maximise production to feed the world' was launched. The tension between the mainstream story so often plastered across the news that food production needs to increase to ensure that everybody has enough to eat and the experience of farmers who struggle to stay in business as they produce ever bigger mountains of food for ever lower prices is an interesting one and one that we will return to.

The divide between the producer and the consumer, now strictly delineated groups, was widening again from the physical to the cultural. One had an experience of over abundance, the other believed the tale of scarcity.

The package of high yielding cereals, machinery and agrichemicals that spread about the world was known as the Green Revolution and has had tremendously far reaching effects.

By now the enclosures had spilled out of Britain, across America and then on to vast tracts of the rest of the world as wealthier nations declared ownership of foreign lands and their right to make money from them in the form of luxury export crops. The Green Revolution was marketed as a way in which the rural, landless poor of developing nations could be fed, however, this narrative also bulldozed the other side of the argument, the idea that it was not a lack of food that lead to hunger in such areas but a lack of access to land on which to grow food or decent wages with which to buy it. The hidden hand of the Green Revolution package removed any hope of just land reform by collapsing local economies with influxes of cheap grain which drove people off the land and into cities. It allowed the developed nations to continue to enjoy cheap local labour and to occupy the best agricultural lands around the world allowing companies to produce luxury export crops cheaply whilst maintaining a positive narrative of 'feeding the world' at the same time.

Industrial Livestock

Livestock were the other main pillar of the mixed farm but with arable production having been freed from them they were also freed from grazing the rest periods of arable rotations.

Our grazers including cattle, sheep, ponies and geese, unlike us, have large and convoluted digestive tracts equipped with an army of plant digesting microbes. This means that they can grow and thrive on plant leaves alone and has historically meant that they have occupied a central role in British agriculture. Our grazers wandered the lands that were not well suited to growing annuals, such as the hills and wetlands, to gather energy from native plants and convert it into meat and milk. Our omnivorous livestock, including pigs and poultry, were the perfect waste bins searching out and gobbling down any scraps of food to convert into meat and eggs. In short, historically there has been a very strong link between our lives, our livestock and our wild landscapes all having evolved alongside one another for thousands of years.

The advances in grain production, as well as the headache associated with the edible mountain of surplus that they caused, pushed agribusiness to look for an ever bigger rug under which to sweep the problem and livestock, cut loose from their traditional roles, fitted the bill well. Poultry and pigs lead the way guzzling down tonnes of surplus grain and by-products from food processing aided by dairy cows, fish and increasingly beef cattle and sheep. These new feeds needed new breeds that could make best use of them and, starting in the 1800s, selective breeding for fast growth rates combined with attributes that consumers desired took off. Pigs became fattier, then leaner once more, then longer bodied to carry more bacon and then white skinned to provide a cleaner looking carcass. Poultry became bigger and bigger, their breasts swelled and they grew faster, they too became white. Dairy cows' udders enlarged and then their legs grew longer to carry their vast bags safely above the ground and beef cattle became bigger and leaner. The livestock, just like the cereals, had been cut away from the systems of which they were a part and left exposed to be used as a lucrative conversion tool in a new system of agribusiness. Unlike the cereals, however, that were still tied to soils in which to grow, livestock could be severed completely and moved indoors into incredibly efficient factories that ship in grain at one end, collect manure at the other end and churn out a colossal quantity of cheap meat in between. Over the last 50 years or so average production of livestock has risen by around 80% due to selective breeding and dietary advances.

Interestingly, our diets too have shifted to accommodate the animals that can consume the most grain. From 1890 to 1990 the global poultry population increased by 1525%, a vast proportion of this being attributable to chicken. Pigs with a similar set of grain guzzling traits increased by 951% and cattle, who can fatten on grain and by-products and produce our prodigious supply of milk have increased by 406% reported in the book 'Something New Under the Sun'. Since the 90s this trend has continued.

Poultry have been the trail blazers for this process and today the number one meat eaten in Britain is chicken. Just as wheat eclipsed oats and rye, chicken has eclipsed our historical birds of choice, pigeon and goose.

Pigeons are the oldest domesticated birds in the world and possibly lived alongside our hunter gatherer forebears in the caves of Mesopotamia tens of thousands of years ago and it's difficult to overstate just how important that relationship has been to us over the years.

When I was growing up in the wake of the miners strikes and collapse of the coal industry in the Midlands a flock of pigeons circling the pointed roofs of the red brick houses was a common site. A retired miner, stick in hand, flat cap on and lurcher at his heel, once told me that when he had been down the pit the only thing that made the pressing dark bearable was his birds and watching them fly free above his garden when he came home. Although pigeon meat had already fallen from favour it is in my lifetime that the pigeon has gone from being a much admired bird to vermin.

Pigeons historically found food for themselves in the form of small seeds and fruits that they gathered from across the landscape and returned to their roosts with. The dovecots to which they returned provided a central place from which people collected the valuable manure, meat and eggs that the birds produced from foraged resources. These yields were drawn from the bird's connection to a diverse landscape and range of habitats around the farm. Their value was in their connections. In today's divided world, focused on single crop productivity, pigeons lost out.

Geese too have suffered a similar fate. They supported small scale agriculture across Europe for 2000 years but, just like the pigeon, their strength lay in their role in diverse and connected landscapes worked by peasant farmers where the geese happily grazed alongside cattle, ponies and sheep. With their thrifty nature and abundant meat and fat reserves, geese were used for everything from butter on bread and oil in cooking to lubricants and seasonal feasts underpinning much of poor rural life in Britain. Historically, geese grazed as a family over summer and were ready for slaughter just as autumn approached with a lean, grass fed bird slaughtered to celebrate Michaelmas, the harvest festival that falls on September 29th, and a heavier and fattier bird that had benefited from some gleaned grain, dropped amongst the stubble at harvest, retained to celebrate the winter solstice.

Today wheat and chicken are ubiquitous and we have all but forgotten about the rye, pigeons or geese that sustained all but the very richest for thousands of years. Today wheat is unavoidable and encountered in bread, pies, pastries, pizza bases, pasta, couscous, semolina and noodles and chicken eggs and meat are everyday fair. The changing tastes for cereal and poultry demonstrate the shifts that have occurred in our agricultural systems as people have turned from eating what a wilder, interdependent Britain could offer up from its tangle of relationships to eating the surplus individuals produced from an industrial food supply chain dedicated to productivity.

Over the last 50 years or so, despite being totally dependent upon the earth, the industrial food system seems to have managed to view itself as existing independently. It seems to assume that the landscape is just a setting and the soil just a substrate on which the marvellous inventions of man can play out. We seem to have conned ourselves into thinking that we can disassemble the world into units and maximise the productivity of those units with no, or negligible, side effects. The modern food system has at its core the belief of civilisation being fundamentally separate from, and dominant over, nature. And of nature being at best a pretty thing to be watched from the kitchen window and at worst the enemy of production. This divided view has crippled our food system, and our world.

Chapter 2

Ripples of Change

It is staggering to think of the scale of change that has occurred in Britain and even more staggering to think of how rapidly the changes of the last few hundred years have stacked up. For thousands of years people were deeply rooted in landscapes, collaborating with cattle to till and sow fields of rye or oats and gathering up what pigeons had foraged for. Very rapidly we have severed these connections, departed from our landscapes and changed our diets to accommodate agribusiness. We now see wheat and chicken as dietary staples and drink milk year round with little thought as to why these things are so freely available. Our baselines have shifted and the water through which we swim profoundly altered.

Although these changes have largely gone unnoticed the rapid alterations over the 1800s did raise more than a few eyebrows sparking the development of both organic agriculture and the conservation movement in protest, groups that would go on to play pivotal roles in starting to reconnect the damaged relationships of our world.

The Organic Opposition

The speed with which land had been separated from people and people transformed from participants in landscapes to consumers in cities didn't go unnoticed. Neither did the stripping down of complex landscapes into linear chains linking a farm, via a factory to a commodity crop sold to the highest bidder around the world. Many people were left warily observing and one such man was Sir Albert Howard, a British botanist and agricultural adviser stationed in India over the early 1900s. Howard realised that the local Indian farmers suffered very few pest or weed problems in their diverse and interconnected farm systems compared to the crops being grown in Britain at the time using the latest agribusiness methods. The discoveries of Liebig's Law of the Minimum had grown into a general assumption that soil was quite simply matter that allowed plants to stand upright and that all plant 'food' could be added in chemical form. Howard realised that the relationship between plants and soil was far more complex than this and developed a method of farming based on composting and returning wastes to the land to build organic matter and nourish soils. Howard firmly believed that all health was rooted in the soil and that instead of combating pest or disease problems whole system health should be improved from the bottom up by building connection and balance once more between plants, soil and livestock. Others joined in with his opposition to chemical based farming techniques including Lord Walter Northbourne, who first coined the term 'organic' in his book 'Look to the Land' published in 1940, Lady Eve Balfour who initiated a long term trial on her farm in England comparing organic to chemical practises and the American, Jerome Rodale, who started publishing the 'Organic Farming and Gardening Magazine'. They all were starting to turn the corner from division back towards connection but only looked within agricultural systems. How farms sat in the wider, and wilder, landscape was not their concern. The wild existed beyond the farm gate and was the focus of another group of radical new thinkers.

Conservation of Fragments

The speed with which nature had succumbed to man's advancing technology over the 1800s exposed its fragility and by the close of the 19th century the seeds of the National Trust, the RSPB and the Wildlife Trusts had been sown in response. Nature was no longer the encircling wild that had pressed for so long and so menacingly upon human pools of civilisation; it was now broken, tamed and vulnerable. British artists started to depict the beauty of nature conjuring into being a fondness of the wild and an awareness of its fragility within the public consciousness. This new view of nature as fragile and exposed to the power and brutality of industrial humans, combined with its distance from the daily lives of city populations, probably contributed in no small part to a shift in how people started to perceive the relationship between people and the wild. The wild was now beautiful and something to be treasured and preserved, against the ravages of man.

Across the pond, in North America, a similar and interlinked process was underway. As the Americans had raced across the New World the wilderness had retreated before them and its fragility, too, had been exposed. George Perkins Marsh was one of the first to write, in the 1860s, of the impact of human activity that he had seen on the natural world and to call for greater measures to protect it. The opening of the world's first national park, Yellowstone, in 1872 followed closely both to preserve wild lands and to make money out of them as the new railway networks linked up wild vistas with wealthy city audiences launching the idea of ecotourism for the first time. A rash of preserves followed across America and Canada and by the 1930s there was a strong emphasis on the preservation of wilderness for its own sake spearheaded initially by the likes of John Muir and Henry David Thoreau and taken further by Aldo Leopold. Thoreau experienced wild nature as sublime and famously wrote:

When I consider that the nobler animal have been exterminated here; the cougar, the panther, lynx, wolverine, wolf, bear, moose, deer, the beaver, the turkey and so forth and so forth, I cannot but feel as if I lived in a tamed and, as it were, emasculated country... I should not like to think that some demigod had come before me and picked out some of the best of the stars. I wish to know an entire heaven and an entire earth.

This desire to know an entire and powerful nature led eventually to the passage of the Wilderness Act in 1964 which identified wilderness as:

> *a place where the land remains untrammelled, where man is a visitor who does not remain.*

In Britain we lacked anything that could come even close to wilderness by American standards and when we set up our own National Parks, the Lake District, Peak District, Snowdonia and Dartmoor, in 1951 they were not an attempt to preserve wild nature for its own sake but were instead an attempt to preserve a cultural heritage. They were echoes of a rose tinted farming past, a pleasant nostalgia into which city workers could slip in daydreams of lives long past. The national parks, and the many other nature reserves dotted across and between them, were chosen primarily to protect specific *things*, a species that was in decline here, a cultural identify that was threatened there or a building left as a relic of a bygone age. They were focused on *individuals* that we perceived as valuable but made little attempt to see them within their context, the relationships that they formed with the wider world. A field might be preserved for its rare orchids but the yearly grazing by the old cattle breeds of a local farmer that had created the pasture in the first place tended to be overlooked. The orchids were valued and conserved, their relationship to the cattle and the farmer often not considered.

Conservation and preservation were born from the same mindset of division that had largely created the threats to our wildlife in the first place. Agribusiness had carved our land up into blocks of monoculture production and conservationists staked a claim to left over areas, fighting to preserve nature reserves like time capsules sealed away from the external world. Our food systems had been enclosed for hundreds, and in some cases nearly a thousand, years at this stage and now our wild lands were heading towards a similar fate.

Whilst maintaining reserves as isolated bubbles seemed a good and meaningful way of conserving our wildlife and honouring our heritage time has highlighted its inadequacies as species have slipped one after another from our landscapes. The emerging science of ecology was soon to shine some light on why an approach based on division and segregation did not seem to be working.

The Trouble with Fragments and Science

We walked to school every day, past the lightning tree, along the green that hosted the fair each year and under the limes that marched proudly down either side of the street. Over summer bright green caterpillars would rain from these trees seeking earth and finding only tarmac and my mum and I would scoop them up and return them to the boughs. Sometimes we would take one home to hatch out. Clinging to its heart shaped leaves, white flecked and pointy tailed the hawk moth caterpillar would almost disappear, its cylindrical dark green droppings the only sign of its presence. One year we watched as the moth hauled itself from its twitching cocoon then trembled as it inflated its wings on my hand. Its crumpled mess transformed into a perfectly patterned, velvety smooth being sensing the air before it with long filamentous antennae. Then, it flew off.

I grew up in a family, in a context, of sustainability. We walked rather than drove, camped above British beaches over summer, litter picked the woods and ate food from an organic veg box. My life was sheltered from the big problems of the world but very much tied to the smaller concerns of everyday life. Doors were to be closed, fridges kept shut, lights switched off and fast food never eaten. Diet was high on the list of ways in which my parents defined, and acted upon, their morality. We were either vegan or vegetarian and I continued to be for many years. I had read 'Gaia: A New Look at Life on Earth', closely followed by 'The Revenge of Gaia' when I was 18 which had cemented my resolve to become a conservationist. I had opted to study Countryside Management at a local agricultural college and diligently learned what I was told. The college was set in the grounds of an old stately home and housed many students on many different courses including those wishing to go on to be vets, others studying animal management, equine studies, horticulture, arboriculture and floristry. The college also taught agriculture at various levels. It never occurred to me at the time but over the two years that I was there we never shared a single lesson with the agriculture students. We had separate classes, classrooms, teachers, locker rooms and tools. We went on separate field trips and ate at separate tables. We never even spoke to them as far as I can remember. There was an iron wall between us and them that did not exist with learners on other courses, we shared business studies with floristry and some practicals with animal management and equine students. Conservation and agriculture, however, shared no common ground.

After college I enrolled on a BSc in marine biology and coastal ecology, the closest course that I could find to freshwater ecology. My degree was in many ways brilliant and I volunteered at the weekends with a local conservation group, carrying on something that I had been very keen on in college. Both of these experiences, my degree and voluntary work, however, had the slightly unintended effect of completely switching me off from conservation.

I remember very clearly, when I was still at college, working alongside a group of seasoned conservationists on a wetland reserve. The reserve was home to a wide array of life who liked water and not woodland and so the woodland was controlled. We spent days cutting young willow and hauling it across the drier islands to a great bonfire up on the ridge to burn. Some willow was left but most was sacrificed on the giant pyre to the god of management. As we finished an area a more senior conservationist would appear, kitted up to the eyeballs in PPE, with a knapsack sprayer of herbicide on her back. A quick blast on each of the stumps ensured that the trees did not come back next year. As we continued cutting we got ever nearer the river and on the last day we were clearing sections of its bank. Again out came the knapsack sprayer dousing each stump right down to the water's edge. Glyphosate, more commonly known as Roundup, is not something that should have anything to do with conservation and certainly not something that should be used near water in my opinion and yet here we were, spraying away. That was the first nail in the coffin for my desire to become a freshwater ecologist.

The second nail occurred as I sat in a lecture theatre in my third year at university. The lecture series was on the threats to marine systems, we had covered overfishing and how difficult it is to regulate and now we were on to climate change. The atmosphere was becoming increasingly laden with carbon dioxide which was being transferred rapidly into our seas. Our seas are the biggest sink for carbon dioxide that we have and the ease with which the gas slips into water allows the earth system to buffer itself from fluctuations. The sea can only take so much, however. In water carbon dioxide turns into carbonic acid and, of course, acid tends to dissolve lime based things like mollusc shells, the skins of sharks and coral reefs. The lecturer stood at the front of a class of maybe 100 twenty year olds and said that the coral reefs were not going to be about for much longer so we should fly to the Great Barrier Reef to see them while we could. He also added that we should go quickly because not only were the reefs threatened with being dissolved from the bottom up by the rising horizon of acidic water but inexperienced divers kept bumping into them and chipping bits off, or else deliberately taking bits home as souvenirs. The roundabout nature of advising people to do something that he had spent his career highlighting the negative effects of (and had come to realise was probably terminal) summarized for me the complete backwardness of conservation and the hopelessness of the science that supposedly backed it up. That was the second nail.

The third nail was my dissertation project. I studied the threat responses of Notonecta, or underwater boatmen as they are more commonly called, the small bugs that hang on the underside of the meniscus of pools. A part of my project was to collect boatmen and crush them up in a mortar and pestle and add their body broth to a tank with a living boatman in it and monitor how many attempts to leave or hide the bug made. This was science, and it was what informed conservation. Tiny incremental additions of knowledge that assumed a mechanical world in which we could, and had the right to, play god. I felt that science conducted in this manner was never going to be capable of embracing the complexity of the world and conservation was never going to protect our biosphere from the threats that it faced. Both science and conservation seemed to me to be safe, insulated worlds in which a person could pour a life into one small question or action. There was no big picture, no real acknowledgement of the world as a whole. What was the point of pursuing conservation, tinkering with tiny areas of land and adding to the big problems that faced the world at the same time? What did it matter if a reserve was wooded or wet when climate change would descend on it regardless? Why should I go and see something if my seeing it would bring it closer to never being seen again? Why spend a life contributing to the real threats to the world to regulate the insignificant ones? It only took three nails, I was out.

I turned my back on conservation and science quickly and in response to something that I would only come to understand much later. I had sensed, rather than understood, that conservation was a small world divided from the larger world on which it depended. I had also had my first true taste of modern science, of something that still attempts to objectively observe and prescribe without entering into relationship with what it measures. It is a dead and cold world view and one that I rejected. I went in search of a larger, interrelated world. Where people and nature fed into one another. I sought out the third pillar of our disconnected relationship to land. I entered organic agriculture.

Teetering Towards the Whole

Over the 1940s rockets had hurtled through our atmosphere to gather piecemeal photographs of the earth's skin but it was not until the 60s that the first colour images of the earth shot from the moon reached the ground. These images, where earth hung like a gleaming jewel in a sea of darkness, made for the first time the planet feel small, isolated, fragile and breathtakingly beautiful in its ocean of black absence. It was a human population looking ever outwards to space that also led James Lovelock, tasked by NASA to look for life on other planets, to look more closely at life on our own. Lovelock was joined in the 70s by radical biologist Lynn Margulis and together they pieced together the Gaia Theory. Gaia Theory is the idea that the Earth functions as a whole, or rather is an 'emergent property of the interaction between organisms' that allows it to remain stable enough to support life.

When I was at university one of our lecturers told us about two species of sandhoppers that live side by side in the strand line on beaches. Sandhoppers are the small cousins of shrimps and prawns that live in the rotting rolls of seaweed deposited at high tide. When only one of the species of sandhopper was present it munched doggedly away at the undersides of the tangled mats but when both species were present they worked faster, they helped each other out to process more seaweed and provide more sandhopper bodies to feed the eager plovers and pipits foraging for them. This increase in productivity is an 'emergent property'. Emergent properties are things that come into being unexpectedly from a seemingly disjointed collection of parts that do not possess the property individually but when combined, do. Two sandhopper species working together can achieve more than either can on its own, they produce a system of decomposition that is more than the sum of its parts.

The Gaia Theory suggests that the myriad species on earth and all of their infinite interactions result in emergent properties that stabilize the whole earth system, it too is more than the sum of its parts. Global temperature, ocean salinity and atmospheric content are just three of the many aspects of the earth system suggested to be mediated by the earth 'super-organism', Gaia. These interactions have evolved over hundreds of millions of years, spiralling one into another to make an incredibly strong system that possesses a remarkable level of stability and resilience. The highly connected nature of the system, however, means that when one aspect of it is altered, something removed or a relationship severed, the whole structure morphs because of it. A good example of the regulatory ability of Gaia, and also its fundamental underpinning of the rest of life, can be seen in our atmosphere.

Over 2.5 billion years ago the early earth had very little oxygen in its atmosphere but the advent of photosynthesis was set to change this. Over the following billion years the level of atmospheric oxygen rose and rose as algae harnessed sunlight, carbon dioxide and water to grow and released oxygen as a by-product, staining rocks red as the new oxygen gushed over their surface and reacted with the exposed metals. The curious thing about the rising atmospheric oxygen levels is that they did not rise indefinitely. In fact they rose to a level very similar to that found today (21%) and then they stopped. Over the last billion years oxygen levels in our atmosphere have not varied by more than a few percent. It was in this steady atmosphere that life bloomed and diversified, able to take full advantage of the predictable availability of oxygen. But why is the level of oxygen in our atmosphere so stable and why has it been so stable for such a long time and through so many turbulent ages of the earth? The answer appears to lie in the older forms of life, those that had existed in a world before oxygen and were driven into hiding by it. The microbes that now dwell within animals digestive systems or in the deep layers of silent decay, swaddled by water, where no oxygen can reach them. They use an older form of metabolism often converting hydrogen and carbon dioxide into methane and water. This methane is released as the bubbles that break out of bogs or the belches of cows. Methane has recently found fame within the realms of climate science where it has gained the reputation of being one of our greenhouse gasses, we will come back to some of the issues surrounding it in this context later. Within Gaia theory, however, it has been suggested that this methane is acting to regulate the oxygen level of the planet by voluntarily combining with oxygen in the atmosphere to make carbon dioxide once more and so completing the cycle. The archaea that live in the pockets that remain of an oxygen free world seem to act to stabilize the oxygen permeated one that succeeded theirs.

The fundamental message of the Gaia Theory is that life and land are bound up more tightly than we can ever imagine and that a change in any area is sensed throughout the rest of the world, affecting it in mysterious and complex ways.

The fascinating and beautiful whole that is our planet started to materialize in the public's consciousness and slowly crept into how we see farms and conservation reserves and how we study science.

The Science of Systems

Although the mainstream scientific narrative has been dominated by an idea of the world as fundamentally divided and mechanistic, there has been another view of the world that is just as old; a systems view. In ancient Greece, Aristotle pondered on the interconnection of the world and much later Leonardo da Vinci imagined the universe as a dynamic, interconnected whole subject to constant transformation.

These minor systems views persisted but it was not until relatively recently that they started to gain traction in the mainstream. King Oscar II of Sweden and Norway was celebrating his 60th birthday and decided to mark the occasion by setting a prize for the person who could solve the so called 'three-body problem'. Isaac Newton's law of universal gravitation works really well when it comes to predicting how two bodies in space will act on one another. It doesn't work, however, when you add a third body. Henri Poincare won King Oscars prize but he didn't actually solve the problem, instead what he produced was the first mathematical description of chaos, published later in 1890. What Poincare demonstrated was that it was the starting positions of the planets that determined what happened next; the system was predictable but the tiniest difference in its starting configuration ended in massively different places.

The study of dynamical systems had taken its first real step but it would have to wait until a coffee break in 1961 for the next. Edward Lorenz was working on weather predictions, feeding numbers into equations that simulated aspects of weather including wind speed and temperature. He set the computer off running and went to get a mug of coffee. When he returned, he realised that the computer was churning out some very unexpected results. Lorenz tracked the problem back to a rounding error. To save time he had rounded one of the numbers from six decimal places to three and this infinitesimally small difference in the starting state of the system had been magnified up to give very different predications of weather for the next two months. Lorenz remarked that the system was sufficiently sensitive that the flap of a butterfly might result in a tornado on the other side of the world, and it became known as the butterfly effect.

Again, mathematics had shown that there was predictability in the universe but the tiniest unaccounted for alteration could lead to wildly different results. This in effect meant that the universe could never be predicted because every tiny movement, every molecule, would have to be accounted for and programmed into the model at the beginning to produce accurate results. The universe might very well be clockwork but it was of a level of complexity that was way beyond our discovering.

And the trouble didn't stop with rounding errors, it continued to the very smallest level of our universe. Werner Heisenberg, in a letter to his friend Wolfgang Pauli, expressed a budding idea that had been growing in his mind; that it was not possible to know the momentum and the exact position of a particle at the same time. The closer one got to pinning down one of these variables the further one got from grasping the second. This later became known as the 'uncertainty principle' and was published in an early form in 1927.

This on its own throws up some difficulties combined with a universe that can only be hoped to be predicted by pinning down the exact starting positions of everything, everywhere. What is more is that Heisenberg also realised that the very act of observing the particle in question altered how it behaved. Not only was it impossible to know the movement and position of the building blocks of our universe, but watching them changed what they were!

There is a fundamental interconnectedness, complexity and uncertainty in our world. Not only was the world not a metaphorical clock, it was not mechanical. It wasn't just non-linear, it was chaos.

Although given the name chaos, and totally unpredictable

in many ways, there are repeated patterns that emerge in our world. The weather might rain ten days from now or it might be bright sun, or mottled cloud. But it won't be bacon, or the sea, or absent. It is unpredictable in the details but the region that it stays in; the patterns and processes that play out as weather; are similar. And this pattern holds true across a wide array of complex systems, from the action of the human brain to ecosystems. Exactly what happens is unpredictable but the general area, the region in which these processes play out, is restricted. This can be plotted and mapped out by scientists and patterns emerge as a result. These patterns are often fractals and they can be thought of as existing in between worlds; they are process and interconnectivity made visible. Fractals are shapes like the repeating patterns that appear in leaf veins, the outlines of deciduous trees and river tributaries. We also see fractals in unfurling bracken fronds, seashells and sheep's horns. Fractal patterns often repeat throughout nature on different scales and when we zoom into or out of the picture the same patterns emerge again and again. These fractal patterns are an emergent property of chaos.

In fractal geometry there are no discreet jumps from a 1D line to 2D box and then to a 3D cube. Fractals are said to occupy intermediate spaces, a line that forms a shape, never crossing but ultimately filling the page, as a lot of Celtic designs do, is no longer 1D, and it's also not fully 2D either, it exists somewhere in between. And likewise the surface area of the intestines cannot be said to be 2D but neither is it fully 3D. Fractals reveal to us that there are no individuals, there are no discreet entities, and everything blends into everything else. We can draw arbitrary boxes around certain elements of a whole, like a whirlpool in a stream, and call it an individual. But zoom out and the whirlpool is embedded within an ever changing movement of water. Zoom out again and the stream is embedded within a shifting network of tributaries. Zoom out again and it is just one braided river flowing away from a mountain range drained by a network of waterways. Keep on zooming out and the whole universe is one interconnected, ever moving process. This fractal, interconnected world is dynamic, it is ever moving and it is interlinked.

The crucial insights from scientists studying complexity and chaos have been that there are no 'things' in the world, just strands of interconnected 'process'. And that these processes can never be untangled from one another or predicted. They can appear to be predictable in that they will repeatedly form similar patterns, like a stream that forms a standing wave; the wave looks the same from one moment to the next but the water from which it is composed is ever moving; and there is always a chance that a tiny disturbance will send the patterns spiralling off into new areas.

Part II

Connecting, Rewilding and Regenerating our Present

Chapter 3

Nature's Patterns

Almost 100 years ago a budding ecologist named Charles Elton set sail for the arctic. He was one of only 20 students and faculty members carrying on a tradition at Oxford of scientific expeditions to far flung places. The young Elton landed on Bear Island, a small island in the Barents Sea above Norway, where he and the slightly more senior botanist, Vincent Summerhayes, had just one week to survey the 12 mile long island. After a week of tirelessly capturing, wrapping, labelling and packaging a vast number of the islands life forms, most of which were flies or spring tails, the team boarded once more and pushed on north to the larger islands of the archipelago of Svalbard. The main islands provided more in the way of life to sample and after a fruitful two months Elton had over 30 boxes of neatly preserved specimens to load back onto the ship for the return journey to England.

Once back in Oxford it was up to Elton and Summerhayes to categorize what they had found. Luckily for Elton at between 74 and 81 degrees north life is spread very thinly meaning that there are only so many species present to be found and recorded. The paucity of life, which at first disappointing to the young zoologist, soon turned out to be a blessing. Elton realised that it offered the opportunity to map all of the interactions between the species. He realised that, in such harsh conditions, food was hard to come by and nutrients, passing from one animal to the next, bound all life together.

In 1923 Elton and Summerhayes published a paper mapping for the first time these 'food chains' as they called them and how they melded with one another into 'food cycles'. They started in the sea, from which most nutrients originated in the form of dead sea creatures washed ashore or dung deposited by birds that fed off marine fish, and traced the interactions across the islands. Plants absorbed the riches jettisoned by the sea with flies and mites feeding off the plants. Spiders ate the flies, ptarmigans ate the spiders and arctic foxes ate the ptarmigans. Elton realised that all of life is bound together by what it eats and what eats it.

The Base of the Web

Today ecologists still see nutrients and energy as the foundation of all ecosystems and continue to map and learn from how they move through food webs. The vast majority of energy on earth comes from the sun and is captured by an army of plants that cover the earth and phytoplankton that float in the surface waters. Phytoplankton living in the sun drenched surface waters of our seas capture what nutrients is available and fix it with the help of the sun into their small, delicate bodies. These tiny reservoirs of life are eaten by a passing whale or else die and drift out of the light down to rest on the dark sea bed where a legion of bizarre decomposers dwell ready to process the rain of life into their own bodies.

If the light and nutrients had been captured on land it is more likely that it would have been pulled into a tangle of grasses or drawn up into a mighty tree where it may have stayed for a thousand years before falling in a broken limb to the ground to take another hundred years to decompose. If the nutrients and light had instead been absorbed by a sphagnum moss growing in an acid bog on a high plateau it may have remained there in the stem of a moss long dead, pickled in an anoxic sauerkraut of peat, for many thousands of years.

Just as nutrients and energy become the captives of ecosystems other nutrients become escaped felons. Nutrients are forever leaching out of one system into another as leaves fall into a stream, rock fragments blow from a windy ridge to collect in the valley below or tiny particles find their way into the atmosphere and are deposited once more as fog blows over networks of leaves.

Animals too enter this dynamic dance, feeding off vegetation, or each other, hijacking the acquired energy and nutrients of other individuals and using them for their own purposes. Animals can take a bite of vegetation that may otherwise have remained for hundreds of years in a woodland and transport it in their guts as they migrate across a continent or else salmon feeding at sea can move nutrients back into rivers and then into the great northern forests via the digestive tracts of bears. In short, the whole world is united by energy and nutrients flowing through it in massively complex ways. All food webs feed into one another forming what could be described as a food globe.

On Bear Island, Elton observed that most nutrients originated from the sea and it was this that limited the productivity of the island as a whole. A similar thing holds true for the vast majority of British habitats but they are generally curtailed by how many plants they can support. Plants, being the main converter of sunlight into biomass, are the gate keepers to ecosystem productivity. The more energy plants can capture and turn into nutritious leaves, fruits and seeds the more animals of all kinds can be supported. The quantity of plant biomass is crucial as the bedrock of food webs.

We tend to think of Britain as being a bit of an overcast and gloomy island, but sunlight is actually very rarely the limiting factor to plant productivity. More often it is the leaf surface area that a plant has at its disposal over which it can absorb sunlight that limits its productivity. Leaf surface area can be measured for different plants and the quantity of leaf surface area to ground area compared. At a ratio of about 1:6 (where there are 6 metres square of leaf surface area to every 1 metre square of ground surface area) almost all of the available light is captured and used up. This ratio of leaf area to ground area is known as a leaf area index (LAI) with a 6:1 ratio expressed as a LAI of 6. Grasses and pasture generally have a LAI of between 2 and 3 and temperate deciduous woodland one of between 5 and 6. Deciduous woods therefore tend to be able to make use of more energy than other temperate ecosystems because a tree can hold many leaves above the ground giving it a larger leaf surface area. One might assume that this should mean that British woodland can grow most rapidly and create the most biomass for animals of all sorts to consume. This is not quite the case, however, whilst a tree can hold a lot of leaves and capture a lot of sunlight it can only do so with the use of a large and strong trunk and branches. This vast scaffolding system owes its strength to the high levels of carbon that it contains but when the tree dies or braches are shed there is a massive glut of carbon rich matter deposited onto the woodland floor which can pose a problem for decomposers.

When building a compost heap old vegetable matter is heaped up with the idea of trying to maximise how rapidly microbes can decompose it so that the matter can be spread back onto the earth to feed the next wave of vegetables. In a compost heap the gardener tends to very carefully load 'greens' and 'browns' on to the heap in layers to promote decomposition. A 'green' is any material that contains a lot of nitrogen, such as lawn clippings or other fleshy growth. A 'brown' is something that contains more carbon, such as cardboard boxes or the tougher stems of plants, for example. In a compost heap the balance of carbon to nitrogen for optimal composting is about 30 to 1, a woodland floor can be as high as 500 to 1 seriously retarding decomposition. And, just like a compost heap, if the nutrients are tied up in plant fragments like trunks that won't break down they are not available to be absorbed by the growing trees next year. Woodlands have abundant leaves with which to capture sunlight but they lack the nitrogen needed to decompose dead matter quickly to fertilise new growth.

If this weren't bad enough woods are also very wasteful with the nitrogen that they do have. In spring, when temperatures rise, bacteria start to actively break down the organic matter on the woodland floor and free nitrogen and other nutrients for absorption. Unfortunately our trees are slow to get going in spring and are in no fit state to absorb the liberated nitrogen which is in danger of washing away in spring showers. Some people have suggested that the carpet of early wildflowers is not just a way for the flowers to avoid the summer shading of the canopy trees but also a mechanism through which they can grab nutrients when it is available and lock it up until they die back and are decomposed later in the year when the trees have fully awakened, passing the nutrients on to them. In a woodland carbon litters the ground but nitrogen is like gold dust.

Nitrogen is actually incredibly abundant, in fact about 78% of the atmosphere is nitrogen, but plants can't access it in this gaseous form. It first needs to be fixed. Only a few families of microbes have the biological equipment necessary to fix nitrogen from the air into a form that plant roots can absorb. Some of these bacteria live free in the soils, migrating around as they please, and others live co-operatively with plants. Certain types of plants can create 5 star accommodation for their microbe allies, such as the legumes that include clover and beans, and the microbes transfer staggering quantities of nitrogen to plant hosts in exchange for accommodation. When these nitrogen fixing plants die their bodies decompose and the load of nitrogen that they had accumulated is released to the rest of the soil food web. There are very few nitrogen fixing plants that grow in woodlands (which is partly why nitrogen is so hard to come by) but in grasslands there is an abundance of them including vetches and clovers. Grasslands have much softer plant tissues, lower in carbon, and an abundance of nitrogen when compared to woodlands allowing them to cycle their nutrients rapidly supporting abundant growth. Grazers can speed this cycle even more by essentially fast tracking decomposition fertilising pastures to provide new and succulent growth. Grasses also have dense mats of roots that are quick to start growing in the spring and late to finish growing in the autumn mopping up any available nutrients over most of the year. This rapid cycling of nutrients allows grasslands to be surprisingly productive but there is a catch, grasses are generally short in stature and unable to support the very high levels of leaf surface area found in wooded systems.

When combined woodlands and grasslands complement one another perfectly; the trees capture the sunlight and the grasslands capture and hold the nitrogen. Together trees and pasture exhibit the emergent property of increased plant biomass and can grow much more rapidly providing far more food for animals than either can alone. Ecologists are increasingly coming to view grassland and woodland as part of the same dynamic dance feeding into one another and enabling each other's growth with their complimentary traits of fixing and securing nutrients and providing scaffolding to acquire the suns energy, benefitting the system as a whole.

Over the years by separating grasslands from woodlands and replacing both with annual cereal production agriculture has robbed itself of a massive quantity of energy that could have been captured by plants; entering our food webs and bolstering our soils once decomposed transforming a deadly greenhouse gas into the base of food webs and fertility. Putting aside what we have subconsciously picked up from living our lives within a divided landscape, it is strange to me that we would ever have assumed that woodlands were one discreet entity and grasslands another.

A Blue Filter

Some systems, primarily freshwater ones, do not rely on nutrients to be cycled or fixed within them but instead function as nets to catch nutrients leaked from other systems. The ever downwards flow of rivers, for example, combined with the fact that they are often not ideal places for either plants or plankton to photosynthesize because of murky water or fast flows, means that they often depend on the productivity of the land from which their waters drain. Rivers take this donated productivity and use it to power their food webs as autumn leaves, branches or the occasional dead sheep wash into them. This brick-a-brack jumble of remains gets caught periodically in dams in the river created by fallen trees, tangles of roots or beavers. In these blockages the matter is snagged and held still to be quietly worked on by swarms of detritivores releasing ever smaller chunks of matter that slip through the blockage and are washed downstream into the filtering mouths of mussels and other creatures.

Seasonally rivers spill out over floodplains or become entangled in networks of lakes, wetlands and backwaters slowing water down. The hither and thither dashing current dies and the turbulent waters, in which plants struggled to grow, slow. The sediment gathered up and bundled along by the frantic river is dropped, allowing plants the clear and still conditions needed in which to photosynthesise. The luscious vegetation further slows the progression of water straining finer particles from the water column and creating clear pools carpeted in freshwater mussels between which myriad invertebrates prowl preyed upon by dragonfly nymphs and water beetles which in turn feed a plethora of birds and fish. Beavers would have glided into these waterways and then doggedly worked away until land melted into water seamlessly. They would once have been responsible for maintaining our streams and rivers, not as one watery thread cut through our landscapes, but as a wet slick oozing down valleys.

In an experiment to determine how beavers learn to build dams Lars Wilsson rescued and reared 4 beaver kits away from the guidance of their parents and put them into a stream to see if they could fathom out what to do. The inexperienced beavers got straight to work building a dam across the flow. When Wilsson placed the beavers in a water free yard and played them the sound of running water they did exactly the same thing. Beavers, it would seem, have an almost OCD reaction to flowing water, if you can hear a trickle; bung it up. The sculptor extraordinaire of our water ways seems evolved to go head to head with gravity.

Wetland soils gather more and more nutrients over successive years of sedimentation and plant growth because, although the water above them is generally well oxygenated, the sediments themselves are not. This means that the bacteria that can break plant matter down the fastest, the ones that need oxygen to fuel their work, are excluded and the job is given over to a much less efficient bunch (often the methane producing microbes that we met earlier). The high levels of plant growth combined with the very slow rates of decomposition ensure that wetlands capture and then hold on to a lot of fertility and carbon in the form of steadily decomposing plant remains and sediment.

Much like woodlands and grasslands our watery ecosystems act as one. Streams, rivers, lakes and wetlands are all part of the same giant blue filter. Year after year, sediments are gathered up by rivers and deposited in wetlands providing enormous levels of fertility producing a staggering base on which vast food webs of fish and wildfowl are built. When farmers initially drained many of our wetlands they profited from this store of fertility but also severed the connection between streams and their wetland sediment traps. By confining water into thin rivers away from their floodplains and wetlands we have failed to filter nutrients from our waterways. Instead of building the super rich soils of the future, sediment now slips through our fingers and out to sea where it settles on, and suffocates, marine life.

The interplay of grasslands and woodlands supporting one another's productivity and wetlands acting as their nutrient bank allowed rich and varied tapestries of life to pulse across landscapes capturing and holding gigatonnes of carbon, building fertility and providing a truly massive quantity of plant biomass on which vast food webs could be built.

Natural Climate Solutions

In 2017 a group of scientists and conservationists introduced a new term, Natural Climate Solutions, to describe nature's toolkit to lock carbon away out of the atmosphere. Unsurprisingly the toolkit revolves around woodlands, grasslands and wetlands and is based on supporting plant growth. The carbon based sugars that plants create during photosynthesis can be swapped like currency in the soil food web for nutrients that the plant needs. Up to 70% of the sugar that a plant makes is shipped down and out through the plants roots to be swapped for nitrogen with microbes or exchanged for phosphorus with fungi, among other things. What is important, from a natural climate solutions perspective at least, is the vast quantity of carbon that enters soils on this 'liquid carbon pathway'.

The other way that carbon enters our soils, the solid pathway as it were, is through decomposition. When a fragment of plant or animal matter falls to the ground it will be swarmed by decomposers who will break down and release the nutrients and energy that it contains. Eventually the decomposers will transform it into a stable version of soil carbon known to gardeners as organic matter, or humus. This humus is not only our carbon store but also our water store and each 1% increase in organic matter in the soil equates to an additional 20,000 gallons of water storage per acre. Our soils hold the key to both capturing some of the excess carbon from the atmosphere and buffering against some of the effects of climate change, such as the enhanced likelihood of droughts and floods.

The more plant biomass that grows across a landscape the more carbon enters our soils and unsurprisingly it is when woods, grasslands and waterways are allowed to interact as one that the greatest stores of carbon can be accumulated.

A mature woodland of older trees can capture on average about 6 tonnes of carbon per hectare per year and a grassland can achieve about 2 or 3 tonnes per ha per year. Rapidly growing scrub, however, with its dynamic mix of sunlight capturing trees and nitrogen fixing herbs, can pull down about 20 tonnes of carbon per hectare per year and fertile, lowland wetlands are capable of capturing similar quantities.

Recreating a network of wetlands linked to our rivers would have other carbon benefits by filtering the water of soil particles and other detritus that washes from the land and over our off shore carbon accumulating and storing habitats. Salt marshes and sea grass meadows, along with other habitats on our continental shelf, capture and store carbon as well as stabilizing our coastlines but they can only do so effectively if they are not covered by the silt discharged from canalised rivers.

If our island was cloaked in a mosaic of connected wood, scrub and grassland threaded with wetlands a truly vast quantity of carbon could be pulled out of the atmosphere and stored.

Climbing the Tiny Food Web

When an organism eats plant matter its digestive system tries to liberate as much of the nutrients and energy it contains as possible. This is by far the most difficult step in food webs because it is far harder to break down plant matter and reassemble it into animal matter than it is to break down one type of animal matter (a grasshopper, for example) and reassemble it into another type of animal (a chicken). In order to break down plant matter some pretty specialist digestive equipment is needed and microbes are the only organisms to possess it. Animals that can host a lot of microbes within their guts can break plant matter down and those that can't, struggle.

In general invertebrates are good at extracting nutrients from plants being equipped with microbe supporting guts and, once they have released the energy from plant matter, they use it very sparingly not generally bothering with heating their bodies. They invest their energy instead in growth or reproduction.

Birds and mammals on the other hand, especially large mammals, tend to be incredibly frivolous with how they spend their liberated energy using vast quantities of it to heat their bodies to high temperatures year round. There is comparatively very little energy left for them to invest in growth or producing offspring. For comparison, a caterpillar might be able to convert about 30% of the energy in plant matter into its own biomass whereas a mammal would do well to achieve 1.5%. The detritivores, including insects and many fungi, working slowly through a sea of partially digested material can often convert about 90% of the energy available into their own bodies or those of their offspring. The energy traded between invertebrates is often what keeps ecosystems ticking over, passing plant matter back to decomposers and on to an army of miniature predators. The invertebrate world of caterpillar grazers and wolf spider predators is incredibly diverse, dynamic and important to the functioning of larger ecosystems. A recent study, conducted in a forest in North America, highlights the importance of the less visible portions of a food web. The researchers found that the population of red-backed salamanders was responsible for five times the amount of energy being made available to secondary predators than all of the species of bird combined.

In agricultural systems we are often guilty of overlooking the importance of the invertebrate food web. We tend to direct plant energy straight into the most frivolous spenders, our livestock, and so end up shipping most of the energy captured by plants straight back out of the system as waste heat retaining very little as animal bodies or fertility. The creatures that can actually hold energy and nutrients within systems to allow them to pass through more transformations and support a greater diversity of life, the invertebrates, are massively reduced and frequently labelled as pests. In effect we put a bottle neck in agro ecosystems by attempting to deny energy moving through a potentially much more productive invertebrate route and focussing instead on a very restricted large mammal or poultry route.

The Movers and the Shakers

Prior to Elton's expedition north ecologists had believed that species existed at relatively static population levels. Elton, however, realised that their populations fluctuated over time and believed they were pushed towards balance by the rest of the web of life; their predators and prey. When one organism's numbers increased its predator multiplied soon afterwards taking advantage of the abundant food supply. Munching its way through the surplus prey items predator numbers increased until the prey had been exhausted. With less prey to eat, predator numbers fell once more until both predator and prey returned to a balanced background level.

This idea rested on the assumption that food webs were constructed from the 'bottom up' with scarce plants determining the numbers of herbivores and the numbers of herbivores limiting the numbers of carnivores. Prior to the 50s this idea was widely accepted but why then was the world green? If herbivore number rose to eat what vegetation was available why were any plants left over? These questions were picked up in the 50s and

60s by three researchers, Nelson Hairston, Lawrence Slobodkin and Fred Smith (known as HSS) who proposed that there must be a top down element to the regulation, that predators must be having an effect on prey numbers for there to be any untouched greenery. Robert Paine was one of Fred Smith's students and it was in search of an answer to this question that led him, in 1963, to start peeling starfish off the rocky coast at Mukkaw bay and throwing them back into the sea.

Starfish prey on mussels, amongst other things, and Paine's removal of them from the food web twice a month started to have some surprising effects. Young mussels started to fill in gaps on the rock, pushing different species of algae, sponges, chitons, limpets, barnacles and anemones out. By removing the mussel eating starfish, Paine had removed the predator that kept the mussels in check and they rampaged over the rocky surface replacing a diverse array of life with a blanket of mussels. Paine coined the term 'keystone species' to describe species, such as the starfish, that have a disproportionately high level of power in ecosystems relative to how many of them there are.

In 1971 Paine teamed up with a student, James Estes, who was studying the sea otter populations of the Aleutian Islands in Alaska and another keystone became clear. Estes was keen to understand the relationship between the sea otters and the vast kelp forests that they lived in just below the waves. On some islands the kelp and sea otters flourished and on others there was only bare rock and sea urchins. Sea urchins are voracious consumers of algae such as kelp, rasping it from the rock surface, and sea otters are voracious consumers of urchins. The trade in otter furs over the 1800s had resulted in a massive decrease in the otters' range and in how many mouths there were eating sea urchins. In the absence of the sea otters the urchins had gnawed away at the kelp forests slowly converting them into urchin barrens; bare expanses of picked clean rock. Sea otters didn't so much live in kelp forests as create them by consuming sea urchins just as starfish created diverse rocky shores by consuming mussels.

Sea otters are another keystone species and Paine coined another term 'trophic cascade' to describe what happens when they are removed. A trophic cascade is a rippling from the top level of a food chain right down to its roots. Since then keystone species and trophic cascades have been identified around the world. Some are mediated by predators, as the mussels and urchins are, and some are driven by herbivores.

African elephants are the largest animals on earth and can eat a truly vast quantity of food; up to 1000 pounds of vegetation every day can vanish down their throats. They sculpt their environment by ripping up trees, digging watering holes in the earth and trampling saplings into the soil. Their feeding clears away the forests but it also disperses tree seeds carried in their guts to new locations, stimulating copses to germinate elsewhere. Elephants, in other words, keep trees on the move, consuming them in some areas and summoning them from the grassland in others. Unfortunately the power and gargantuan appetites of elephants also means that they come into conflict with humans. They destroy crops, knock down houses and periodically injure people and that, combined with the ivory that they carry, has resulted in drastic oscillations in their population level. This has meant that elephants have, over the years, been removed from certain areas and confined in high densities to others and their relationship of renewal with woodlands has been pushed out of balance. In areas where elephants have been removed trees have spread and, without the elephant populations to keep up with them, they have proliferated turning diverse savannah into sparse woodland. As grasses are replaced by trees the complex fauna of wildebeest, antelopes and zebra as well as the lions and hyenas that hunt them melts away. In areas where the elephant populations have been imprisoned, and sometimes maintained in artificially high number to attract tourists, they have stripped the land back to its bare bones of sand and rock. Once more the staggering diversity of life that thrives in the elephants wake trickles away. The disruption of the relationship between elephants and trees sends ripples right through the ecosystem lowering population levels of a diverse array of plants and animals adapted to the landscape created in the wake of stable herds and trees.

As scientists have mapped food web relationships they have accumulated a long list of species that play unusually powerful roles in creating landscapes across the world. From Antarctic krill to prairie dogs, from sharks to bees certain species sculpt and steady ecosystems but once their populations are destabilized, those relationships alter and landscapes change and those changes radiate out into adjacent landscapes, rippling across the world.

We will come back to these keystone species when we discuss rewilding later in this book but for now we will focus on the dance that plays out between herbivores, predators and plants. Herbivores graze plants, browsing back grasses, herbs and trees and fertilising the ground with their manure stimulating further growth. Left to their own devices herbivores tend to strip plant life back to its bare bones as high populations of deer have done across the Scottish hills and elephants have done in areas of Africa. Predators mediate herbivore numbers giving plants the chance to grow up and out of the reach of their munching mouths allowing a tapestry of trees and grasslands to establish. The enormous quantities of plant biomass produced by the association of woodlands and grasslands are only made possible through the browsing of herbivores, which opens up areas of pasture, and hunting by predators, which protects pools of saplings. Complete and dynamic food webs are needed to create a whole landscape and whole landscapes are required to support dynamic food webs. They are all part of the same whole. Grasslands, woodlands, rivers, wetlands, herbivores and predators; they all create one another and are all part of the same entity.

Scientists studying food webs have realised that the food web does not start at the first mouth in the sequence and end with the dung of the highest level predator, it spills beyond digestive tracts, and it permeates landscapes; knits them together, or unravels them. Feedback loops tie the highest to the lowest strata of the web of life together over vast distances; sometimes across the entire globe. The simple act of an elephant consuming the plants around it sculpts an ecosystem which feeds back into the patterns of rain that pulse across a landscape, the levels of nutrients in the soil and the quantity of carbon in our atmosphere.

How Species Connect and Systems Evolve

As a young man Charles Darwin left his studies in theology at Cambridge to take up a 5 year voyage upon HMS Beagle as the ship's naturalist. Darwin was an avid collector of plants, animals and observations and his years aboard the Beagle left him with a wealth of specimens and careful sketches to pore over back in England.

Some of Darwin's better known collections were his Galapagos finches. Each of the 13 species of finch was unique and possessed a different type of bill including one for hard nuts, one for smaller seeds, one for insects and one for cacti. As Darwin studied his finds an idea started to materialise in his mind, these species had all come from a common ancestor that had changed to fit in with its environment. Darwin understood that there were limited resources in the world, such as food, and that it was spread between many eager mouths. He realised that some mouths were better than others at accessing certain food types and that those individuals would end up being better fed and leaving more, and healthier, offspring behind them. The offspring who inherited the mouth shape best suited to the diet on offer would also be more successful and over time species would change to suit their environment. Darwin proposed that each of the finch species had evolved through natural selection, a bill well suited to its specific food source. Although Darwin suggested at the time that there was likely to be a whole host of different mechanisms pulling and pushing species, sculpting them to thrive within specific environments, a lot of research has tended to focus on the role of competition in a resource scarce environment. The idea of competition and resource scarcity was rife in 1800s England, forming the backdrop to Darwin's life as he wrote 'On the Origin of Species', and he was alive at a time long before HSS and Robert Paine started to challenge the assumption of bottom up ecosystem regulation (although Darwin did ponder the issue himself). This still leaves us with a quandary however; if resources need to be in short supply to drive evolution how does evolution work in an ecosystem that is regulated by predators from the top down? In other words, in a green world what drives grazers to evolve?

Although much focus has been placed on the role of competition for limited resources, researchers, including Roberto Cazzolla Gatti, are finding examples of evolution in low competition environments. What may be happening is that species are evolving towards rich areas of resources instead of away from scarcity; it may not always be a lack of available resources that drives evolution but an array of possibilities instead.

Evolution can be pictured, not as species desperately vying for what little there is available in a bottom up world, but as species moving towards resource richness in a top down one. This might lead over time to more species using the same or similar resources transforming one thing, grass, for example, into zebras, wildebeest, gemsbok or Thomson's gazelles; creating even more resources for other species to utilise. Together a diversity of grazers return more fertility to the grassland fuelling more growth of an array of different plants adapted to different nutrient levels and depositing fertility in the form of different types of droppings, each of which can be utilized by different species of dung beetle, for example. All of that diversity in the form of different styles of munching mouths, dung shapes, beetles and fertility can support even more types of life including big cats or dogs, small predators and insect eating birds, mammals or reptiles. Different scavengers, from insects to hyenas, can also use different carcasses at different times and can be eaten in turn by even more organisms or plants whose seeds are adapted to different digestive tracts can bloom over different areas. At each stage an abundance of resource options allow life to flow into them transforming energy into yet more resource options and building diversity over time.

Scientists have found that increasing diversity leads to an increase in complexity of the relationships found in systems. For example, if we had a field in which clover and ryegrass were grazed by cattle, we have a small number of relationships; the relationship between the grass and the nitrogen fixing clover, the relationship between the grass and the cattle and the clover and the cattle. If we consider an African savannah on the other hand there are millions of relationships between all of the different plants whose roots reach to different depths in the soil and support different microbes, between all of the invertebrates, the birds, the rodents, the large grazers, the predators, the parasites, the fungi and pathogens. The system as a whole is a network of relationships binding it together as one and, just like numerous elastic bands strung between the species, those relationships give the system strength and flexibility. They make it stable. The more stable a system becomes the more of these relationships can form as species evolve to rely on one another's stable provision of resources, adding once again to diversity and enhancing stability. The process is self reinforcing; cooperation, diversity and stability progress together.

As diversity and stability increase in a system so too does productivity with more transformations of more resources allowing more biomass to be sustained. The system's level of resilience rises too. Resilience is a measure of how well a system returns to 'normal' after a disruption or disturbance. If a very simple system, a tilled field, for example, is exposed to cold winds it may kill off the delicate shoots of the grower's crop or deposit a layer of dandelion seed across the field. The area would turn from a neat field of beans into a sea of yellow flowers. The bean field is not stable, it struggles to resist the winds disturbance that transforms it into a dandelion field. The field is also not resilient; without external assistance it will not spontaneously return to being a bean field.

A low level of stability can be pictured as a ball on a hill, one tap (the strong wind) and the ball will roll off the hill and down to the valley floor (the field of dandelions). The ball would have to be hit very hard to get it to roll back up the valley side and onto the top of the hill once more which represents the low level of resilience of the system.

If the strong winds had hit a plantation of one or two species of conifer with a ground cover of rough scrub the wind might knock sections of trees down. Young conifers could be ready to spring up to replace their felled parents or birch seed might blow in and get the upper hand, moving the wood from a conifer plantation towards birch woodland. If the winds were strong and persistent enough they might create a patch large enough that grasses or heath could develop and the system might move towards an open glade in that area. This system is more stable because it would need stronger winds to fell an area of established trees. It is also more resilient because the area is more likely to return to being a conifer woodland, or woodland of some description, than it is to switch to grassland. This type of system can be pictured as a ball in a depression on a hill. The ball will not move out of its depression if only gently tapped;

the system is more stable. The ball will also be more inclined to roll back to its initial starting point if it is moved a little, but not sufficiently to push it over the lip of its depression; the system is more resilient. The system still has limits, however, knock it too hard and the ball will bounce over the edge of its depression and roll off down the hill into a different stable state.

The most resilient type of ecosystem is a very diverse one, one that holds within it a wide array of tree, herb, grass and wetland plants as well as an abundance of animals all ready and waiting as adults or as seeds in the soil to awaken when the time is right. If a hurricane hit such a diverse system it could rip sections of trees and scrub out but a similar set of species would soon grow back creeping in from the edges or springing from seeds buried in the soil to repair the damage. The system as a whole would still resemble a diverse scrubland and would still function as one. This type of system can be pictured as a ball in a deep trough; an incredible level of energy would be needed (like a tsunami or land slide) to hit the ball hard enough to smash it out of its rut and into a place from which it could not slide back.

Chaologists, those who study chaos, would say that systems always hold within them order and chaos at the same time. They show us an ordered, apparently predictable face and a chaotic, disorderly face and have the capacity for both at all times. Although the ball on the hill analogy is a useful one to picture stability and resilience, it is just an analogy and no more accurate than Descartes' clockwork universe. Systems are always moving and evolving and although they can appear to us to be quite stable, quite stationary like a ball on a hill, they are never static. Ecologists call systems 'stable' but perhaps they would be better thought of as being in a process of remaining similar, of showing us an ordered face. At some point, however, the system will alter and show its chaos face of rapid transformation and apparent turmoil before returning to order once more. Ecologists call this return to order resilience but again it is not a return to a fixed thing or defined place, it is just another strand of process. Change really is the only constant.

In wild systems millennia of interaction have yielded highly diverse systems with a truly vast array of species which are producing a ridiculously complex web of interactions that coats the globe like spiders' webs weaving together all life and land through almost limitless interconnectivity. This web of interconnections gives ecosystems their stability, productivity and resilience and, just like a ball bound together with infinite small elastic ties, when it is thrown against a wall its elasticity allows it to flex, morph, bounce back and return to its original shape very quickly. So far we have talked about how species link together in ecosystems to form coherent wholes, in the next section we are going to take a look at how these wholes flex and shift through time.

Links Through Time

For a long time, since the 1800s in fact, scientists believed that ecosystems progressed from a bare start of rock or scree through pioneer communities of herbs towards a climax community of woodland. This idea was very convincing at first but over the years scientists started to realise that the sequence very rarely seemed to play out as it should. Unexpected things kept happening that would get in the way of the sequence of succession and over time scientists started to see the role of chance as more and more important to the process of ecosystem development. Today the importance of chance occurrences, or disturbances, is widely accepted.

Systems are disturbed by any number of things including grazing, rooting, flooding or even a jay storing its food for spring, anything that disrupts the functioning of an ecosystem or 'knocks the ball'. As time passes stressors can accumulate in a system such as a bank of gravel that gradually builds higher and higher in a stream as pebbles are added one by one over the years. At some point the gravel bank will reach a critical size and alter the water flow, pushing the stream into a new course. The steady accumulation of gravel is an example of a threshold disturbance; something that builds until a point is reached that forces the ecosystem to respond and alter. Other events are known as sledgehammer disturbances such as a tree that falls into the river disrupting its flow all in an instant. Mixtures of these types of disturbances are always passing through systems, always interacting with them whether we can see them or not and at some point the stable pattern will alter, the system will change rapidly and unpredictably revealing its chaos face.

The scale, duration and intensity of disturbances massively influence what effect they have especially on the diversity of the system with very high and very low levels of disturbance linked to lowered biodiversity in general. Intermediate levels of disturbance tend to support higher levels of diversity by maintaining a more complex habitat. For example, if a fire blazed across a landscape every week there would be very few plants or animals that could inhabit the ecosystem. On the other hand, if an ecosystem is never disturbed, if a woodland grows up and nothing ever topples a tree it will shade out its understory and its diversity will fall. It is an ecosystem with an intermediate level of disturbance that supports the most diverse array of life by periodically opening up opportunities for species to move in.

As disturbances clear small patches in an ecosystem of their previous occupants the door is left wide open for whichever species can get there first to take up residence. Some species are very well equipped to jump from patch to patch opportunistically (like many of our annual weeds and pests), others bide their time as seeds in the soil or saplings waiting to be triggered to grow by an influx of light. Some are very good at rapidly scrambling in from the edges of a disturbed patch to sweep over the fresh ground. Each time a new patch opens up there is an opportunity for different species to enter ecosystems.

One thing that has become quite apparent from a variety of studies, notably by Daniel Simberloff in 1969, when he fumigated whole mangrove islands to study which species came back, is that the identity of the species that colonise a patch is not all that important and that as systems become more diverse, productive, stable and resilient the identities of the species present matter less and less. As long as enough species reach the patch to fill all of the roles needed such as decomposers, herbivores and predators, for example, their actual identities are not all that important. In fact an ecosystem that looks similar and functions in much the same way can be made up of a massive array of different species. The mangrove swamp, in Simberloff's experiments, is the thing that is persistent not which species actually form it.

In the vast majority of ecosystems there is a continual level of disturbance that allows species to move from patch to patch, the habitat in affect becomes a network made up of areas at different stages of recovery after a disturbance and colonized by potentially different sets of species. For example, many butterflies used to skip between the coupes coppiced by woodsmen for charcoal, feeding off the abundance of flowering herbs that sprang up once the canopy of coppice had been removed.

For both conservationists and farmers, the patchwork dance is an opportunity to either support or hinder different species as desired. Conservationists manage fragments of habitat to support certain sets of species, recreating ancient disturbances to maintain set populations. However, nature works in broad brush stokes across a global tapestry with grazed pastures, scrublands, wetlands, woodlands and wood pasture rippling across our island infinitely flowing onwards. When a section of that dance is cut out, its edges defined and a certain step repeated over and over again it ceases to be a dance at all and becomes a slightly odd staccato movement instead, its vitality gone. This happens when conservationists freeze reserves in time, endlessly coppicing a small area to preserve a species of butterfly, for example, or when farmers till the same patch of ground each year to grow annual cereals. It isn't the application of a disturbance to a place that allows a species to remain there, it is the continual flexing of landscapes over time that opens up niches here, there and everywhere in which species can participate in the dance. The disturbance happens in a place but it is connected to everything else and every other disturbance that has ever happened; scale is of prime importance and it is what conservation has always historically lacked. It is also what rewilding and regenerative agriculture can both offer.

The dance of disturbance is a fascinating one and one that we are only now beginning to appreciate.

Memories of Interdependence

Disturbances and interactions can become so evolutionarily interwoven over time that they are difficult to separate. It becomes almost imposable in fact to pick out any individuals, species or habitats; they are all one and the same. As the famous naturalist John Muir said:

> When we try to pick out anything by itself, we find it hitched to everything else in the universe.

There is no individual, no separation between animal, vegetable or mineral, there are also no separate habitats; landscapes make life and life makes landscapes; they are one and the same. Beavers blend land and water, grazers blend woodland and grassland. Cattle do not find themselves in a field any more than a beaver finds itself in a wetland or we find ourselves in a town, life collaborates with landscapes to form themselves in a moment in time. Song birds do not live in a scrubby patch but they form the scrub as they feed on seeds and perch and defecate on branches, again in a moment in time.

You can remove boar from an island and then reintroduce them hundreds of years later but it won't be the same island, or the same boar. The boar had a memory of the habitat, where to feed, where to farrow, and the habitat had a memory of the boar. Its snuffling allowed shorter lived annuals to flourish which littered the ground with their seeds building a memory within the earth of how to respond should the boar come again. Life and landscape have history, in other words species and habitats remember their disturbances, their connections. Memories play out across our island from the resistances to diseases that each of us carries in our blood equipping us to our home and its microbial peculiarities to the migratory routes of birds returning from Africa each year guided by the magnetic pull of their ancestral seat. When soils are stripped of their microbes and seeds they forget their erstwhile allies that healed their hurts and when species are removed from their landscapes they forget their patterns of movement that facilitated their, and the landscapes', existence in the first place. The dialogue between place and life is like a story sung from one generation to the next immortalising the history of their relationship, a new individual in a different place may to us appear the same but the stability that history provided will be missing. It will be perched atop a landscape, not something that grew naturally from it and whose ancestors had brought it into being, as a whole.

Humans too are a part of this dance, we have co-created our land for generations and continue to do so, although the ways in which we interact have become muddled by ideas about them. Ecology has embraced complexity and offers us a profoundly different view of our world, as an amazingly complex whole of which we can be a beneficial part if we so choose. Conservation and agriculture are two ways in which we are aware of our participation in this dance and we will be discuss them below.

Chapter 4

Rewilding Conservation

Cores, Corridors and Keystones

As the environmental movement of the 60s got underway the debates surrounding how to best protect nature heated up. Wildlife reserves, which had existed in many different guises for over 100 years at this point, were drawn into sharper focus when in 1967 Robert MacArthur and Edward Wilson published 'The Theory of Island Biogeography'. MacArthur and Wilson's ideas roughly boiled down to a central concept; animal and plant populations are maintained by individuals breeding and new blood migrating in from other regions. When populations dwelt on islands the further away and smaller the island was the fewer species would find it to colonize it and those that did would be more prone to extinction because their population sizes would be smaller and replenished less frequently by migration from the mainland. The closer islands were to the mainland and the larger they were, the more species would find them and the longer they would persist, achieving higher population levels and benefitting from the more individuals migrating in from the mainland.

This idea, of islands amidst the sea, was quickly applied to reserves amidst farmland and has been incredibly influential to conservation ever since. Researchers dispersed to some of the largest reserves in the world in an attempt to ascertain if they were big enough to prevent extinctions from occurring within them and they found, very convincingly, that they were not. Although extinction rate decreases as reserve size increases extinctions do still occur at a measurable rate.

In natural systems the distribution of individuals is not even, it is clumped into sub-populations and sometimes those clumps can go extinct. Extinction is a natural process and an everyday occurrence within subpopulations. In order to maintain a species in an area the subpopulation must be constantly bolstered by new blood washing in from other subpopulations. Migration between populations is key to sustaining them and in order to support migration, landscapes need a certain level of connectivity. It is no good having a massive island if it has no connection to the mainland. When we acknowledge the dynamic, clumped patterns of a species across a landscape the focus shifts from trying to maintain a large core reserve to linking that reserve up with other reserves across a landscape.

When compared to the efforts that the British had made when establishing their reserves it started to become clear as to why our island continued to leak species. All of our reserves were far too small and were often in a sea of fairly hostile habitat far away from another patch of similar habitat. In many cases a reserve being meticulously managed for the remaining pair of spotted flycatchers or other species of conservation concern had already failed, the networks connecting them to other flycatchers were not there. Habitat loss and fragmentation became the key to understanding conservation worldwide but waited for another 30 years or so for a further piece of the puzzle to be added.

Michael Soulé grew up just outside San Diego in the chaparral covered hills. Coyotes regularly slunk about his house and preyed on his pet cats pursuing some back in doors and causing others to disappear without a trace. Many years later Soulé was working as a lecturer at the University of California studying the coastal sage-scrub remnants left as the city sprawled ever outwards. Developers had chopped the habitat up, selling it off piecemeal for clusters of large new houses. Some fragments of steep sided land became caught up in this tangle of progression, looking for all the world like native sage-scrub they clung to their gullies encased in a bubble of humanity. With the people arrived more domestic cats who preyed, as Soulé's had done, throughout their landscape. Soulé noticed that within the captured lands the cats prowled free of fear of any coyotes and slaughtered the native bird population whilst any who ventured out into wild and untethered lands beyond soon fell afoul of the native carnivore. What Michael Soulé had found was a terrestrial version of Robert Paines starfish, the coyotes were regulating the sage-scrub ecosystem from the top down. When the coyotes were present they kept the domestic cats, and other mesolevel predators, in line; but when they were removed the lower level predators boomed decimating the native bird population.

On the other side of America, nearly 2500 miles away, a man called Reed Noss was pursuing similar ideas, linking up habitats to facilitate movement of the Florida panther and the Florida black bear. The two joined forces and proposed, in 1998, that 'rewilding' as it had now been termed, was a conservation method based on cores, corridors and, the final piece of the rewilding puzzle, carnivores.

Wolves had been hunted across much of their range in the United States and by the 1920s they had been pushed out of Yellowstone National Park. The removal of wolves 70 years previously had, completely by accident, created a perfect long running experiment into what happens when a high level predator is removed from an ecosystem. A researcher named William Ripple, intrigued by what he may find, set about surveying the age of aspen within the park. He concluded that all but 5% of the trees present had germinated before 1920 meaning that the vast majority of Yellowstone's aspen was over 70 years old. The same was true of the stream side willows and cottonwoods. The date for the last major recruitment events of these trees within the park was from a time just before the final wolves were removed. In 1995 wolves were reintroduced and the same researchers, among others, tracked what effects their reintroduction had recording an upswing in aspen, willow and cottonwood growth.

Gradually a picture emerged. In the absence of wolves the grazers of Yellowstone had bred, multiplying to numbers far greater than the vegetation could support, and their abundance had repressed the emergence of tree saplings. When the wolves were reintroduced they started to prey on the grazers, lowering red deer numbers by half, but they also affected how the grazers behaved. The grazers retreated from certain regions, such as the valley heads and steep slopes, where they were unlikely to escape a wolf ambush and stuck instead to wider and more open ground. These 'landscapes of fear' as they have been termed allow some areas of habitat to go relatively under grazed, resulting in trees soaring upwards along the sides of the valleys and gorges, and other areas to be kept open. The establishment of trees and scrub on steeper lands helps to bind and stabilize the soil with deep roots which in turn benefit the streams that drain to gather on the valley floors.

Middle level predators such as foxes also breed more freely and roam more widely in the absence of wolves, hunting smaller prey instead of scavenging on the wolves leftovers. Once the wolves returned they reversed this trend too, coyote numbers fell by 39%, resulting in song bird and rodent numbers increasing as the coyotes and foxes feasted on other foods.

Once the scrub had regenerated beavers spread into some of the valley waterways increasing from one to 12 colonies in the Lamar valley by 2009. Eating the vegetation and felling the regenerating trees to form their dams and lodges they flooded vast areas and created new networks of wetlands and grazing lawns. This explosion of trees and water allowed more species to move in including fish, reptiles, amphibians and even more song birds roosting and feeding in the new scrub. Bears too came to feed on the berry rich plants spreading the fruit seeds further up into the hills in their faeces.

The unintentional Yellowstone experiment showed quite clearly the power of whole systems, complete with all of the species that had evolved to be there, to heal themselves. The Yellowstone wolves caught the imagination of the world and rewilding took off.

As a predominantly American born concept the reserves that the initial attention had focused on were the large expanses of American wilderness, preserved as areas in which natural processes could play out, Yellowstone being one of them. The idea of wilderness became synonymous with the idea of the core of the reserve, a kernel of wild, self willed habitat beyond the grasp of humans and where human will held no sway. Carnivores were also very helpful in indicating which areas came up to scratch because they are very intolerant, in general, of disturbance from people. They also range very widely meaning that their habitat has to be sufficiently joined up in order for them to continue to use it. Carnivores then were both seen as a vital element of rewilding and also the indicator of its success.

As rewilding has spread about the world the carnivores have been reclassified into the larger category of keystone species in many places. Carnivores are, of course, often keystone species because a few carnivores can have a large impact upon prey species thereby affecting vegetation structure, but so too are animals like beavers and cassowary. Beavers are classed as keystones because one family can coppice large tracts of woodland, consume vast quantities of plant matter and flood entire valleys maintaining a dynamic wetland which supports numerous other species. Cassowaries also affect the vegetation of their homeland because many native plant species will not grow unless they have been previously consumed by a cassowary.

Although crucial to the maintenance of a habitat, keystone species sometimes make poorer indicators of overall size and connectivity of habitats than carnivores, often having smaller ranges or more accommodating attitudes towards people. The duel strand of carnivores and keystones therefore has persisted.

Rewilding as it stands now promotes the establishment of the three C's and a K; Core kernels of self willed habitat where natural processes can play out unhindered, Corridors of safe passage between cores, Carnivores as very delicate indicators of the health of the habitat and Keystones as vital players in maintaining stable systems.

Rewilding at its heart is about uniting vast swaths of land with all of the species that would have been there without the alterations of man so that complex and stable habitats can once more be forged from the age old interactions between land and life.

Baselines and Beginnings

As rewilding spread to Britain people started to question how it could be applied here, to a land devoid of any large cores of wilderness, any meaningful predators or, in fact, any area of self willed land at all. In other places around the globe two 'baselines' have been identified, 'whole' collections of species that serve as start points from which we judge the degradation of the present and to which we should ideally aspire.

The first baseline, the Pleistocene baseline, identifies the start point as the time period spanning around 2.5 million years ago to the point when *Homo sapiens* first left Africa. This baseline is used in an attempt to understand how the world may currently look if modern humans had never colonised it. The biomass of megafauna currently seen across African reserves may very well be indicative of the biomass that once prowled across Eurasia and the Americas and Pleistocene rewilders often seek to introduce elephants or rhinos back into modern day landscapes.

Perhaps the best known Pleistocene rewilders are Russian scientists Sergey Zimov and his son Nikita who are trying to recreate the Mammoth Steppe that once sprawled across

Siberia, Alaska and Canada. In previous ages short grasses would have dominated this landscape grazed by mammoths, horses, musk ox and bison. In this interglacial, however, in the absence of the megafauna's munching mouths, this frozen landscape has instead been colonised by scrub and the great rolling forests of deep green taiga. The scrub and conifer forests increase how much summer sunlight is captured as heat by this great belt of the earth and then they insulate the ground from the freezing winds of winter. These lands lie over the permafrost, permanently frozen ground a mile deep in places, and the vegetation's heat capturing and insulating traits are increasing the permafrost's rate of thaw in our warming world. The thawing of the permafrost is one of the tipping points in climate change models because as the soils thaw their cargo of frozen organic matter can start to decompose releasing huge quantities of greenhouse gases that have been imprisoned for thousands of years in the ice. The temperature rise caused by the released gases would then allow deeper soil layers to thaw reinforcing the affect and starting a vicious cycle that could push us closer to climate collapse. The Zimovs believe that by restoring the great grazers and browsers who once wandered the steppe the vegetation can be downsized and its heat capturing and insulating properties minimised. Without its insulative blanket the ground will freeze over winter and the greenhouses gasses will remain under lock and key. Although the imported horses, musk ox and bison are browsing the Pleistocene park's landscape back towards grassland they are not the tree topplers that the mammoths once were. The Zimovs keep the landscape moving by recreating the action of mammoths by driving what can best be described as a tank into mature trees to push them over. The Zimovs maintain that for their fairly radical approach to tackling climate change to work in the long run they need to bring back the tree toppler herself, they need to de-extinct the mammoth, something that is not as farfetched as it sounds.

George Church is lead scientist of the Woolly Mammoth Revival Project and is pushing hard at the possibility by subtly 'tweaking' Asian elephant DNA to make them cold hardy, long haired and small eared. A type of ibex known as the Bucardo has already been 'de-extincted', if only for a few short minutes before it died and returned to extinction once more and, bizarre as it sounds, a mammoth revival appears not to be off the cards.

A Pleistocene perspective, whilst not the easiest thing to achieve, does raise some interesting questions surrounding what Britain may once have looked like. There is a general feeling that our island was once cloaked in thick wildwood, that oaks flowed from shore to shore and that a red squirrel could have run from John O'Groats to Land's End, if it had so wished. This idea has come under increasing scrutiny more recently as people have applied a Pleistocene perspective to Britain and concluded that our trees appear to be elephant adapted. The vast majority of our native trees grow back well from harsh defoliation or coppicing and require high levels of sunlight in order to grow well which has lead people to suggest that they are adapted to the gaps and glades that our megafauna would have crafted. In fact of our native trees it is only yew, holly, beech and possibly small leaved lime that can grow up under a canopy of mature trees and these species have never been dominant in our landscapes. When we look at the pollen record we find a great deal of oak, birch, hazel, willow and alder present; species that require light, open landscapes in which to grow and uninterrupted breezes to carry their pollen to another's flowers. Our trees not only seem to be adapted to spread their pollen in a light and breezy landscape, they also seem adapted to disperse their seeds in a similar way. Many species such as ash, birch and willow have light weight seeds equipped for flight on high winds to new areas. Other trees that were common in times past, including hawthorn and blackthorn, put on a sweet scented display of flowers to attract a range of insects flapping through sunlight drenched, flower rich landscapes and produce attractive berries or fruits to be scooped up and moved by birds or mammals.

The pollen record is widely used and fairly accurate but, like all information we gather from the past, it is open to interpretation. Fortunately the abundance of different tree species in our landscapes has also been recorded in another way; by how many species have come to rely on them. Insects show us, in their preferences, which trees they have lived alongside for the longest time and which were the most numerous and dependable. Oak supports 284 species of insect, willow 266 and birch 229, and all of these trees are ones adapted to more open places.

Our other species also seem far more suited to a more open, scrubby, landscape. Of the invertebrates, butterflies are perhaps the most easily recognized and most loved and the vast majority of them occupy open habitats or glades. Even of the handful that are classed as woodland species many require grasses or the plants that grow alongside them, including vetches, birdsfoot trefoil, devils bit scabious or cuckoo flower, to feed their caterpillars. Many other invertebrates back up this story of coevolution with a patchier environment including moths, shield bugs and snails.

Birds too shout the story loudly, literally, from our garden hedges. Many of our garden favourites including blackbirds, thrushes, wrens, robins, dunnocks and various tits need mottled landscapes as do our house birds; starlings, swallows, house martins and sparrows. All of these species need scrub or mature trees in which to nest intermingled with clearings in which to feed.

Beyond our houses and gardens Britain teems with yet more birds adapted to a megafauna maintained island including those that have evolved migratory routes to intercept the rich diversity of life a patchy Britain offers. The influx of summer visitors including cuckoos, flycatchers and nightjars arrive each year seeking the insect bonanza that a patchy and wet island can offer. In autumn another great move occurs as birds flood in to feast on rose hips, hawthorn, rowan, white beam and holly berries. Field fares, redwings and waxwings, who can eat 2 to 3 times their body weight a day in berries, journey specifically to feast in our scrublands.

It is highly likely that Britain's native species have a genetic memory of a land maintained by giants. Even the humble robin is thought to have once followed the vast elephant bulldozers around in search of displaced insects as they follow their gardening substitutes today.

Frans Vera, a Dutch ecologist, has been the mastermind behind a lot of these discoveries. He has focused on the roles that large grazers and browsers such as bison, elk, horse and aurochs would have played in maintaining an open landscape of woodpasture, thickets and glades across Western Europe. Vera and his colleagues have added weight to these arguments at their pioneering site in Holland, Oostvaardersplassen, where they allow Konik ponies, red deer and Heck cattle, the breed that has resulted from extensive back breeding programmes to recreate the aurochs, to develop with the land in a natural manner.

The Oostvaardersplassen is one of many projects that take a loose version of the second of the rewilding baselines as a guide. The Holocene baseline runs from around the time that *Homo sapiens* first exited Africa to the period just before the birth of agriculture. It takes as its stating point a world that had already been heavily impacted by hunter gatherers, most notably in the eradication of the megafaunal grazers, browsers and predators, but that had not yet been impacted by farmers and still possessed large herds of smaller mammals; Aurochs, elk, boar, deer, wolves, bears and lynx.

Many of Britain's current rewilding projects have a loose

interpretation of this baseline including Knepp Wildland in West Sussex. Knepp is a 3,500 acre estate within which the neat parcels of fenced farmland have been merged and nature has been given the opportunity to lead the way. The estate is home to longhorn cattle in lieu of aurochs, Exmoor ponies in lieu of the wild tarpan and a small band of free roaming Tamworth pigs in lieu of boar. Together with deer they have sculpted the area into a patchwork of scrub, wood, grass and water that supports sky rocketing populations of all sorts of species including turtle doves and nightingales.

Both the Pleistocene and Holocene baselines rest on the assumption that nature did perfectly well on its own prior to human intervention. Both baselines assert that by giving nature back her tools, including large and connected spaces and a full complement of species, we can best equip her to meet the challenges of the future including biodiversity loss and climate destabilization. Rewilding seeks to recreate the 'entire earth', longed for by Thoreau, so that we may know it once more and allows that earth to determine her own path once more.

Rewilding also tends to assume, however, that the species that should be here now are those that were here thousands of years ago and that native is a strict stance that a species either conforms to or does not. As Simberlof showed, however, it is not necessarily a species identity that is of importance to ecosystem functioning, it is the role that it plays.

Native or Not?

Charles Elton, who we first met as a young man setting sail for Bear Island, later wrote a book called 'The Ecology of Invasion by Animals and Plants', published in 1958. This book launched a whole section of ecology called invasion ecology which seeks to understand what happens when species arrive in new places. Since then the idea of native and non native species has taken off and today whether a species is native or not can determine how a great many people, including conservationists and rewilders, feel about it.

Generally people agree that if a species *evolved* in a place then it is native. In the case of Britain, where species have been wandering in and out for millions of years, we tend to say if it was hear when the sea rose and cut us off from mainland Europe then it is 'meant' to be here. If it arrived after that, especially if it arrived by some action of people, then it is alien. Our culpability in fact tends to dictate to a large degree how we feel about a species arrival.

When Elton wrote his book in the 50s he was looking back at the 1800s and first half of the 1900s, a time period that had seen a massive upswing in our ability to move things, including species, about the world. Enough species had been moved over a sufficiently long period of time that evidence of the effects of species relocations was everywhere. In his book, Elton noted some very good reasons for keeping non-native species out of ecosystems and there are more than a handful of cautionary tales dotted about the world of what happens when invasions are permitted. These include zebra mussels, a species of filter feeding mollusc that can multiply very rapidly in suitable new habitats and strip the water of native plankton collapsing the food webs built upon it. Cane toads are another well known example, imported into places including Australia to eat the beetles that damage sugar cane plantations, cane toads more often choose to eat native wildlife instead. The toads also produce a toxic substance that predators in their native ecosystems have evolved alongside but to which predators in the areas into which they were released have no defence. There are many other examples of non-native species around the world including rabbits, rats, pigs, starlings, carp, beetles, plants, ants and mosquitoes that have become unwelcome additions to ecosystems. These species cause havoc; they eat those that have evolved no defence to them, are unpalatable to would be predators, kick similar native species out of their homes, take native species' food sources or infect them with new diseases. In general the picture that emerges is that new species in old places can, in some instances, spell disaster.

When we look at these species within the larger context, however, there are very few that are actually troublesome. Of the thousands of species that we have moved around the world and that have become established in new areas very few of them cause problems. Most of them just get on with their lives, slotting quietly into a new food web. There are even some intriguing instances of non native species forging beneficial relationships with native ones. A team, including Edwin Grosholz from the University California, Davis, found that an endangered Californian clapper rail actually benefits from a non native grass hybrid called spartina. The clapper rails, rather ungainly birds that wander salt marshes and tidal pools looking for clams, crabs, mussels and spiders, use the dense stands of the non native grass for nesting.

So how should we respond to invasion? Most ecologists have historically favoured removing non native species, even if they do little or no damage, to keep ecosystems in historical states but from our vantage point of the 21st century we can view things in another light. Species movement is unavoidable, the web of life is eternally flexing and evolving as relationships form and break down. We can also view the actions of people in a more integrated and nuanced way. Members of the genus *Homo* have been influencing the spread of species around the world for potentially hundreds of thousands of years as we have hunted them, burned clearings for them, removed or added them and consumed and created resources alongside them. In fact all species are pushing and pulling all other species within the web of life all of the time and people are no exception. We have undoubtedly started to pull threads in the web far more enthusiastically over the last few hundred years but the mutual movement of species is not an alien thing in and of itself.

I suggest that instead of looking at ecosystems as fixed 'things' that were forged at a certain point in our history and are constructed of certain building blocks we look at them instead as processes. The important aspects of ecosystems are diversity and connectivity so if a species is enhancing either or both of these traits then we could refer to it as 'nativing'; they are in the process of fitting themselves into food webs. If, on the other hand, species are reducing diversity or complexity (or are simply losing connectivity within their food webs) we could refer to them as 'exoticing'; they are destabilizing or leaving ecosystems.

This process-led understanding of ecosystems frees us up to value all species including non-native ones such as little owls, horse chestnuts, evening primroses and common field speedwell for the relationships that they have built as nativing species.

It also casts a new light on our exoticing species. The wildlife that was trapped on our island when the waters rose to flood the channel was suited to a different, cooler world. Some of our species are already on the edge of their range, pushed further up mountains as the climate warms like snowshoe hairs or ever northwards until there is nowhere left to go like the capercaillie. Some of our species will probably have to move north to become someone else's nativing species as they become the exoticing ones here. This is not to say that we should accept the loss of these species when it is a human pressure that is eroding their connections within their landscapes, we should perhaps though start to embrace a less static view of which species should be here and what our island should look like.

It is high time that we abandoned our idea of creating a fixed and perfect Eden and dreamed instead of a dynamic and flexible future. And no species needs to engage in the process of nativing, rebuilding relationship and connection with the ecosystems around it, more than people.

A Native Human

A species of *Homo* has lived in Britain on and off for 500,000 years or more, migrating with the great herds of megafauna as the ice advanced and retreated. The genus *Homo* more than fits the criteria laid down for both a Pleistocene and Holocene rewilding baseline and the tendency of rewilders to substitute extinct ancestors for extant relatives paves the way for us, *Homo sapiens*, to be included in lieu of others of our genus such as the Neanderthals. Modern people also more than fit any currently accepted definition of native wildlife having been present in Britain at the end of the last ice age before the sea rose to form the Channel. People, however, are one species that rewilders seem determined to overlook.

It is our adherence to the perceived divide between people and nature, far more than any scientific principle, which expels humanity from the wild. This divide seems so deep that it appears almost unbridgeable and yet people are unavoidably bound up in the world whether we acknowledge it or not. We are the ones drawing mammoths from the grave, reintroducing species to landscapes and defining their edges. We are intrinsic to the success or failure of rewilding and yet we hold our impact, behind the wheel of a tank or peering through the lens of a microscope, to be separate from the role of every other creature alive. We still feel our role as an objective observer of a mechanical world even though science is highlighting our inherent interconnectivity within a dynamic one. We have embraced connecting trees with grasslands and dry with wetlands, we have even embraced apex predators, but we seem unwilling to traverse the chasm that cleaves people from wild.

For the remainder of this chapter we are going to look more closely at people; what ecological roles have we filled in the past and which of these roles are we filling today? Are we nativing or exoticing? Why do we feel so powerfully that people are not, and can never be, wild?

Footing the Bill

We will start by discussing the only widely accepted form of human connection with wilderness; the bank note. Rewilding tends to rely, to varying extents, on ecotourism to bank role it just as its predecessors, the National Parks including Yellowstone, did over a hundred of years ago. Rewilding has found a powerful ally in ecotourism and together they have been able to take strides that rewilding alone would never have been able to take, or pay for, but there is a danger of the band aid of ecotourism being mistaken for a long term ally of nature. A monetary interchange can never be the sum total of our interaction with life, nor should it be in my opinion.

All around the world ecotourism brings in a lot of money, the wolves of Yellowstone are worth more than $5 million a year, and even the sea eagles on Mull are worth over $2 million a year to the local economy. In fact a recent paper, partly produced by the WWF, attempting to get a handle on the value of global ecotourism put the figure at something like $600 billion a year. This is a much needed boost to local economies and it can go a great way towards helping people to view their wildlife, especially top predators, as an asset instead of a burden but if wolves and bears repopulated their native ranges and if sea eagles really did fish off every coast would people still pay so highly to see them? Would everywhere reap these financial benefits or would the money be dissipated and spread too thinly to maintain local commitment to creatures that could very soon start to get a bit annoying?

Ecotourism is not necessarily the most dependable base on which to rest the future of our wildlife. We only need to take a whistle stop tour around the world to see what happens when the money dries up and dependent local populations get desperate. In Nepal, across Africa and over South America wildlife guides have switched to hunters when conservation projects fail to bring home the bacon. How the local population regards its wildlife is key to preserving it, if the wildlife is a cash cow once it stops earning its keep history shows us that it will most likely be eaten.

If we embrace the idea that wildlife may once again cease to be threatened, which surely must be the goal of all conservation projects, or the other possibility, that in a climate destabilized future people may not have the spare time, money or ability to participate in ecotourism, a longer term plan needs to be sought. People who live hundreds or thousands of miles away will never be as good a protector of wildlife as those who live alongside it.

So to me it seems high time to seek some alternatives. To take a fresh look at how *Homo sapiens* are embedded within the web of life.

In Search of the Wild(er)ness

In his book 'Feral', George Monbiot describes his state of 'ecological boredom' and the relief that rewilded lands can offer for this state. A relief that presumably many people seek as they pursue various ecotourism ventures. Monbiot describes the feeling of coming up against the wild. But what is wild? As the basis of rewilding it is absolutely crucial to have a clear understanding of what 'wild' is but we seldom probe any further into what it might be than our initial assumptions. *Wildness* is also more often than not used interchangeably with *wilderness*, but what do we actually mean by the two terms and is there a difference?

Rewilding is built more or less on the American definition of wilderness given earlier in this book, that wilderness is an:

Area where the earth and its community of life are untrammelled by man, where man himself is a visitor who does not remain'.

It firmly establishes the roles of nature and of people in wild areas. But the Americans didn't simply pick this definition out of thin air. They got it from Descartes. He suggested that everything could be split into two mutually exclusive categories, the immaterial realm of the mind and the physical realm of matter. He believed that there was a 'self' that existed within a world of everything else and by extension that there was a cultured humanity that existed in a surrounding 'wild' and uncultured nature. People generally assumed therefore that the 'wild' was anything that was not crafted by man.

This binary manner of thinking, the assumption that something is one thing because it is not another, is very easy for our brains to process. If something is not on, we assume it to be off, if it is not male, we assume it to be female and if it is not cultured, we assume it to be wild. Although quick and easy this method of assessing the world is open to some fairly serious oversights. Firstly a binary view does not allow for grey areas, that something may be neither on nor off, neither male nor female, neither wild nor cultured. Secondly, when things are grouped into mutually exclusive categories people tend to order them and give them a hierarchy. On is better than off, male better than female and cultivated better than wild, for example. Descartes had broadcast to the world over the 1600s that the self and culture were superior to the surrounding 'other' and as people colonised the new world they carried these views with them, creating civilisation from the wild.

As American wilderness disappeared beneath the plough and the call for preservation went up, another but equally divided view of the world was aired by people including Henry David Thoreau and John Muir. They cherished the 'romantic natures', where one could chance upon the sublime far away from the turmoil of humanity. They still saw a deeply divided world they just swapped the hierarchy around so that the wild was better than the civilised. In both views of the American Wilderness it was not a place for people, it was a place for conquest and improvement or for peace and preservation.

Although viewed as wilderness by both Cartesians and preservationists alike, the lands that the colonists came across were far from untouched by man. They were inhabited by many people and they were worked, in some cases very intensively, but in a very different way from the lands of the old world. Native Americans had managed great swaths of both North and South America for thousands of years guiding the ecosystems of which they were a part to offer up more foods than they would otherwise have done. The Native Americans across North America maintained dynamic vegetation covers with fire, encouraged woodlands to support more edible species and cropped root and clam gardens by the coast. In South America people had built great civilisations from the food generated from lands as diverse as the high mountain plateaus to the rainforest valley floors. Today the Amazon is largely regarded as the last wilderness on earth and it harbours nearly a third of global biodiversity, however, researchers are now uncovering just how 'farmed' it once was. There were 138 edible plant species grown in the Amazon over 6000 years ago and over half of these were trees. These ancient people seem to have cleared the understory of the forest to grow maize, squash, peanuts and chilli peppers in very diverse polycultures before planting up the old gardens to edible tree crops and leaving them to mature. Successive generations of careful tree planting and tending have resulted in the Amazon still possessing 30% more edible tree crops including Brazil nuts, cashews, cacao and guavas than would be expected otherwise. These farmers also incorporated additional nitrogen fixing trees and shrubs and carefully cultivated the soils below their crops, adding plant matter to cool burnt wood to build up incredibly fertile soils known as the Amazonian Darks Earths (ADE). Today, these soils are found below about 10% (and potentially much more) of the forest. The Amazon, just like the vast majority of the rest of the Americas, was defiantly not an area untrammelled by man.

Unfortunately, the zeal for protecting lands in a pristine state lead to the forceful removal of Native Americans from large areas including Yellowstone National Park and then Africans from Kruger and many other areas around the world, sweeping away their presence and their heritage as one. The people who had allied themselves to the landscapes for generations were removed as a damaging and corrosive force on 'wilderness' so that Westerners could engage in the fantasy of a binary world. There are obviously many ways in which people can, do and have in the past embedded themselves into local ecosystems but we should not underestimate the power of our belief in a dualistic world.

Even in Britain today the fire ignited for the rewilding of the uplands, in Wales especially, threatens to remove farmers from our hills and erase their lives, traditions and relationships just as we did across America, Africa and so many other places in the world.

Our adherence to the belief in ecotourism as protection and local populations as threatening is old indeed and we must remember that wilderness became unpeopled because of a value and a definition that was placed on it, not because a virginal state is a prerequisite of the wild.

Wildness, as opposed to wilderness, has been proposed as a concept to acknowledge the continuum instead of the extremes, the connection instead of the divide. As a term, wildness ditches the associations that have stuck to wilderness over its long and uncomfortable history and allows us to step away from a binary thought process that holds one state to be superior to another and one place separate from its neighbour. Wildness moves towards an understanding that it is connection, process and relationship 'spun between people, animals, plants and soils', as Sarah Whatmore and Lorraine Thorne write, that can be wild. That the wild does not need defined edges and the exclusion of people in order to exist and have value. That it can be given its autonomy, its self will, and retain its value across great swathes of land in the company of people. Wildness moves towards seeing the value of natural processes wherever they occur and whichever species are bound up in them, our own included. By moving away from wilderness and towards wildness we can open up opportunities for our existence in the world that go beyond monetary exchanges. We can start to examine where people sit within ecosystems once more, how we integrate ourselves and what connections we can forge.

The Human Role

To my mind there is ample reason to embrace people as part of wild ecosystems; we are one of Britain's native species, we were here (or can fill in for other members of the genus *Homo* that were here) for thousands of years and it is not actually possible for us to be separated from wild systems in the first place. The acceptance of people as part of the wild does, however, raise some interesting questions about our ecological identities.

In 2019 Meredith Root-Bernstein and Richard Ladle started digging into what the ecosystem roles of people might have been and concluded that we, and other members of the *Homo* genus, are keystone omnivores. Whilst the ecological implications of the eradication of grazers, browsers and predators have received quite a lot of consideration over the past few decades the effects of the removal of the omnivores has flown somewhat under the radar. Today we are left with much smaller keystone omnivores, including a handful of pigs and bears as well as ourselves, which are quite possibly just as important as keystone herbivores and carnivores but their roles are much less clearly understood. Root-Bernstein and Ladle suggest that, just as grazers mediate vegetation levels and predators mediate grazer levels, omnivores mediate the system as a whole acting as a sort of damping mechanism. Pigs, bears and humans all share very flexible digestive systems which allows us to switch our food stuff over the year eating whatever is most abundant; if there are plenty of deer we will hunt them but if there is an abundance of apples we will collect those instead. Our overall impact is to gather surplus and in so doing reduce the population spikes of other species helping systems to stay on a more even keel and guiding them towards stability. Modern hunter gatherers, such as the Aleut who live on the island of Sanak off the coast of Alaska, have incredibly detailed knowledge of the abundance of the plants and animals upon which they rely. Researchers from the Santa Fe Institute, lead by Jennifer Dunne, have recorded the massive array of species hunted and foraged for by the Aleut including various shellfish, seaweeds, fish and even sea lions. The Aleut people eat about 25% of the species living on the coast and swimming through coastal waters switching between species in response to their availability. The researchers concluded that the hunting and foraging activity of people had helped to stabilise the ecosystem for over 7000 years by harvesting what was abundant and leaving what was scarce.

Omnivores have another trick up their sleeves; they tend to roam large areas following abundant food sources. Bears will flock to the salmon run and then cluster around fruiting bushes and pigs will move between the riches of forests and wetlands, for example. When omnivores move from one area to another they carry with them fertility and seeds in their digestive tracts, transporting them to new regions and spreading plants as they go. Many hunter gatherers, including foragers in the Kalahari, gather up seeds, fruits and nuts and transport them across the desert to new areas. The intentional, and unintentional, movement of species by people around a landscape is not new, in fact over 20,000 years ago, well before the dawn of agriculture, people moved the northern common cuscus, a long tailed tree dwelling marsupial, from New Guinea to eastern Indonesia and the Solomon Islands.

A final trait that we share with other keystone omnivores including bears and pigs is our tendency to cause disturbances in ecosystems. We all root for food, topple the occasional tree and trample down vegetation where we have chosen to rest for a while. Our patches of disturbance, such as the cleared soil left by the Hadza of Tanzania as they dig for tubers, offer species the opportunity to establish in new regions and help to build diversity into landscapes.

It is very easy to find examples of hunter gatherers filling the role of keystone omnivores in landscapes, harvesting abundance, spreading seeds and causing disturbances, but the role extends to a great many more of us as well. When we garden we are yet again playing a keystone omnivore role; we cut back dominant plants, we sow seeds or plant bushes bought in from other areas (often a garden centre) and we create networks of disturbance as we weed or move plants. It is deeply within us to interact with our world, participating in behaviours that link us with the rest of life, as a keystone omnivore.

Humans are not just keystone species however; we are also ecosystem engineers. The difference between a keystone species and an ecosystem engineer is a subtle one that continues to be widely debated 25 years after the term was first used. It does not help that many keystone species are also ecosystem engineers, including beavers, for example, however, there are some important differences between the two and both terms are useful when it comes to thinking about how people interact with the wild. The difference between keystones and engineers partly stems from how individuals interact with the world around them. Keystone species tend to interact primarily through consuming something and making a by-product as a result. Rhinos consume plant matter and convert it into seed laced dung and wolves consume deer, for example. Ecosystem engineers on the other hand tend to impact the ecosystem around them in other ways. They tend to modify it by their presence like coral polyps that form intricate reefs from generations of discarded calcium carbonate exoskeletons supporting an array of other species. Although this rule is not hard and fast it is a useful starting point when taken with a pinch of salt.

The other strand of the difference between keystones and engineers is how much impact a single individual of the species has. Generally, keystone species have a huge impact on their environment for the number present, there are only about 95 wolves in Yellowstone, for example, but each one packs a punch. Ecosystem engineers tend to have an impact through the sheer numbers of a single species; a coral reef, for example, is the product of many millions of coral forming polyps. Although it can be a somewhat slippery definition thinking of species as ecosystem engineers has helped ecologists work out what is happening in ecosystems and it can help us when thinking about people in landscapes.

We discussed earlier the impact that fire had on our ability to cook food, warm ourselves and burn vegetation away. Controlled burns could be described as perhaps our first steps into our ecosystem engineering role and one that we continued. When the first colonists disembarked from their ships to wander the Americas and then Australia they marvelled at the open landscapes reminiscent of English parklands. These landscapes had been formed through thousands of years of collaboration between people, large grazers and the rest of the web of life. In North America human set fires pushed the forest back promoting grasses that were maintained by deer, bison and elk. The fires also encouraged more nut bearing trees including walnut, hickory, oak and chestnut which fed people as well as wild game. It is highly likely that our engineering, allied to keystone grazers such as bison, created a patchwork that stretched across whole landscapes and within which life evolved.

In 2002 Justin Wright and Clive Jones probed a bit more deeply into the question of whether ecosystem engineers promote diversity in landscapes and in general they found that they did as long as two criteria were met. Firstly, that the engineer creates a patch of a different sort of habitat than that which 'naturally' occurs and secondly that the patch must support species which can use the created habitat type or prefer it to the unmodified background habitat type. Beavers therefore raise biodiversity levels because they create wetlands and pools in otherwise dry areas that then support a wider array of species including amphibians and fish. Coral polyps do a similar thing by transforming an area of open sea into a reef that houses countless other species. The same idea can be applied to the mosaic formed by periodic burning and other human activities in an otherwise undisturbed landscape. When we burn or farm people create patches of different types of habitat within a 'wild' backdrop.

The first colonists of New Guinea started to burn patches of rainforest to stimulate the growth of edible light loving plants and hunt the animals that flocked to feed on them. They created gaps that other species could exploit and so raised the diversity of the area as a whole. The first farmers across China enhanced diversity in a similar way by opening up gaps in alder woodland to create rice paddies and, as the Romans marched across France, they established farms in the forests, locally enriching the soil to support different tree species. Across the earth and through time people have *increased* diversity by engineering small areas of 'difference' in landscapes of monotony allowing a greater biomass and diversity of life to thrive.

Scale, however, is key. For ecosystem engineering to have a positive impact on biodiversity it must create a small area of difference within a larger background habitat. Over the last few hundred years or so the speed and scale of our engineering has increased and we have flipped landscapes inside out, as it were, with small pockets of 'wild' caught within a fabric of disturbance. Our engineering no longer provides a novel pocket within a background of wild stability; the 'wild' has become the novel area in a sea of human engineering.

Our ecosystem engineering has transmuted over the years from burning to farming and the keystone species that we have collaborated with have also shifted from wild to domesticated. We have added and removed keystone species from habitats around the world for thousands of years from the extirpation of wolves and sea otters to the encouragement of bison and cattle. These additions and subtractions of keystone species put us in an enormously powerful position and in 2016 Boris Worm and Robert Paine published a review paper in which they identified people as hyperkeystone species; a species that impacts the distribution of other keystone species. When we dispersed around the world and removed the megafaunal grazers and predators we were removing keystones and when agriculture spread about the world much later we introduced new domesticated ones. Now we are exercising our hyperkeystone ability yet again as rewilders reintroduce wolves and beavers.

People not only removed and added keystone species, we also mimicked them. When we eradicated the megafauna thousands of years ago we took up the burning of the landscapes that had been previously grazed. We disposed of the accumulation of plant matter as the megafauna had done but in a different way. When people arrived in Britain we could not burn away the vegetation so we started to fell and coppice it instead, perhaps very closely replicating the ripping and tearing of elephant tusks. We also scoured the landscape for food and deposited the remains of what we had gathered in middens, concentrating fertility in a few areas as Nepal's rhinos do today in their dunging patches.

Later, when farmers arrived, we began the long process of replacing wild aurochs for domesticated cattle but we also replaced native wolves for ourselves. Stewarding our stock about landscapes, encouraging them to cluster in open areas and avoid tangles of woodland, and preying upon them to regulate population densities, we took up the role of the wolf.

We also learned from, and replicated, the beaver. As hunter gatherers we had made fish traps by diverting or pooling water and as farmers and then industrialised people we constructed ponds along tracks to water stock and horses, dammed rivers behind weirs to harness their power and created giant lakes in the hollows of old quarries and the grounds of stately homes.

In urban environments, that have effectively lost all of their keystone species other than humans, we take our mimicry to even greater heights. Our lawnmowers are stabled in our garages in lieu of our grazers and our hedge trimmers hang above them in place of our browsers. Even our garden ponds, surrounded by neatly clipped plants, recall the work of beavers.

Our impact is not just our own, we have learned off those who have gone before. I believe that we deserve the term 'keystone mimic' to be added to the ecological description of who we are. Keystone omnivores, ecosystem engineers, hyperkeystone species and ecosystem mimics - that is our ecological identity.

Ecosystems Great and Small

People, ecologically speaking, play many roles; we influence other keystone species, we harvest abundance, we sow seeds, we create small disturbances, we mimic long lost species and we modify areas of landscapes to enhance diversity. We are hyperkeystone, engineering, omnivores and all of our roles have the capacity to build dynamism and diversity into landscapes through the food webs in which we engage. We are not, however, only engaging in larger food webs; we are quite possibly food webs ourselves.

Microbes, 10,000 species of them, outnumber human cells 10 to 1 in the human body and contribute 360 times more genes that code for proteins required for our survival than the human genome does! Our bodies are works of collaboration between a descendent of an ape and many millions of microbes and it would be fair to say that we can more closely be described as an ecosystem, complete with its own food web, than as an individual.

Viewing ourselves as an ecosystem is a fantastically weird way of understanding ourselves and yet one that makes total sense when you consider the history of life. For around 3 billion years microbes alone inhabited the earth and got very good at communicating with one another using various chemical signals. When multicellular life evolved, arriving unfashionably late to the party 500 million years ago, it existed within a microbial sea and became a very convenient refuge within which many microbes hid. Over time the larger plants and animals learnt the chemical language that their microbe associates spoke and came to rely on their constant chatter. Chemical signals zipped this way and that and multicellular life eavesdropped on the conversation, gathering information about what was happening in different parts of the organism. Together unicellular and multicellular life coordinated the running of bodies that got larger and larger eventually leaving the sea to creep, run and fly over land.

This shared story was only discovered in the 80s when Jesse Roth and Derek LeRoith uncovered just how similar ancient microbe chemical signals are to today's hormones and other molecules created by our microbe partners and interpreted and relied upon by our brains.

Unsurprisingly there is nowhere in the body that is more microbially active than the intestines where billions of microbes plaster a network of over 100 million neurons creating a fantastically intricate merging of microbe and human. The plethora of chemical signals released by the gut are gathered by our brains and the information shipped up the vagus nerve, a sort of super highway connecting our gut microbiome to our head. Gut microbes digest and make available our food breaking down carbohydrates, fats and proteins and creating a barrage of chemicals in response that allows the body as a whole to gather information on the food consumed and the world outside. Many of the compounds produced are absolutely essential to our survival and include vitamins and chemicals that regulate our immune system, mood and memory including serotonin, 95% of which is created by gut microbes.

When we eat a meal the food becomes the substrate on which our microbes multiply. Just as in other ecosystems, certain species of microbe eat certain foods and the more of those foods there are available the larger the populations of that specific microbe become. Certain microbes also produce certain suites of chemicals and so as their populations fluctuate the levels of the chemicals that they produce also fluctuate. This means that there is a link between our diet and the chemicals in our blood stream including those that influence our mood. In a series of studies conducted on mice researchers tested some of these theories and ascertained that the temperament of the mouse could be altered by changing the mouse's gut biota; normally shy mice could become bold and normally bold mice could become shy. Likewise normally happy mice could be induced to become anxious or depressed and normally anxious mice could be encouraged to relax. What these experiments suggest is that the food that we eat sculpts our internal microbe community which releases different chemical signals to our brains triggering us to be bold, shy, relaxed or depressed.

If this were not enough there is now evidence coming to light that our microbes can communicate with our brain to request the food that they would like! Diverse diets sustaining diverse microbial communities tend to be fairly stable, just like other diverse ecosystems, but when certain restricted foods are eaten in abundance certain species tend to become very common and others can disappear. In these destabilised communities certain microbes seem to be able to 'request' our brains to find certain types of foods, including fatty or sugary foods on which their populations can thrive.

This is still an emerging area of understanding but what seems to be appearing is the meeting point between our internal and external food webs. Our human-microbe chimera consumes the food produced by a landscape which then stimulates the human to seek certain types of food *from* that landscape, and potentially to *engineer* landscapes to generate those foods.

The rabbit hole gets even deeper when we acknowledge the impact that our landscapes have on how we feel. Being in natural landscapes lowers our stress levels, heart rate and blood pressure. It also enhances our mood, inspires feelings of awe and generosity, restores our attention, supports our memory and makes us less anxious and less depressed. Basically, nature has a powerful effect on the brain and there is increasing evidence that this is communicated back to the gut microbiome.

In a series of experiments, again on mice, researchers tested the communication system between the gut and the brain, but this time in reverse. They found that stressed mice have an altered gut microbiome which encourages inflammation. Inflammation then triggers the release of a molecule that enhances the brains susceptibility to anxiety and depression creating a feedback loop between brain and gut.

I am only speculating, but I would image that if our landscapes and lives support us to feel relaxed and confident, and produce a diverse and healthy diet, it is highly likely that our gut microbiome will be supported in its fully diverse state which in turn will steady our emotional responses and trigger us to eat a steady supply of nourishing foods. In other words diverse landscapes are probably likely to encourage us to eat like hyperkeystone, engineering omnivores that participate in dynamic food webs which, in turn, facilitate us to co-create stable ecosystems. If, however, our landscapes and lives cause us stress and anxiety we seem more likely to select a smaller array of less healthy foods which reduce the diversity of our microbiome, triggering feelings of depression and anxiety as well as cravings for restricted food groups. Our desire for high fat and high sugar foods then potentially triggers us to manage our landscapes for the production of high sugar and high fat diets supplied by agribusiness models which further degrade our landscapes and lives.

The relationships that weave us into the rest of life, both within and without, are incredibly complex, deep and as yet very poorly understood. We understand far more about the roles of other keystones species than we do our own and we see with far more clarity the landscapes that they create. We have charted the return of wolves to Yellowstone and witnessed how they draw forests from the grassland, we have seen how elephants create the savannahs that sustain them and how beavers sculpt the wetlands in which they live. Keystone species craft landscapes in which they, and the host of species that have evolved alongside them, flourish. If we are going to shift our roles within ecosystems it requires an overhaul of both our internal and external food webs and how they relate to one another.

The Path from the Sea

How people interact with the rest of life has long been a fascination of mine. It was the central strand that guided me towards conservation and it was the strand that guided me away from it again. Not long after the second nail had been driven into the coffin in which I placed conservation I volunteered on a permaculture site in South Devon. A growing unease was rippling through my mind at that point; people would not leave areas alone for wildlife if it conflicted with what they wanted, and they always seemed to want more of everything. Plymouth sits on a crag of land between two large rivers, the Tamar to the west and the Plym to the east. The Tamar cuts its path, winding between Bodmin and Dartmoor, across the land to empty into Plymouth Sound. By the time the Tamar reaches the sea it sits above a deep limestone gorge hewn from the rock by centuries of moving water. Flowing above sunken cliffs and old shores the Tamar hides the 40m deep scars left by an ice age river that once carried melt water from the glaciers above Wales to the great 'Channel River'. The river Plym shares a somewhat similar history disguising below its surface its own deep ravines and networks of now sunken rivers that once flowed over the cattedown plateau.

The whole of what is now Plymouth Sound was once an ancient limestone valley clad in cold grasslands. Today the sea has claimed the valley for a different set of creatures. Seagrass beds flourish amidst rocky outcrops sheltering juvenile fish and cuttle fish whilst kelp forests bloom further out to sea in the blast from the ocean beyond. A network of sponges and barnacles coat exposed rock edges and the limestone outcrops have become home to cold water reefs with a wide array of anemones, rock dwelling species and sponges covering their surface. These reefs give way to a different type of reef built, not on bedrock but formed from life itself, as you travel to the tops of the gorges and nearly exit the sea; beds of mussels, oysters and cockles.

The ancient depths of Plymouth Sound allow it to support a vast array of life but they also make it a very useful place for people. Plymouth hosts the largest naval base in Western Europe, with HMNB Devonport commanding a whole 4 mile stretch of the water front, and it is also home to Brittany Ferries who ship people back and forth to France year round. These large boats need deep waters in which to sit and the Sound is perfect but conflicts with wildlife are not uncommon.

Plymouth Sound is designated as a special area of conservation (SAC) with specific areas within it also designated as sites of special scientific interest (SSSI), special protection areas (SPA) and marine conservation zones (MCZ). These designations should afford Plymouth Sound a very high degree of protection and are some of the strongest designations that we have for wildlife habitats in Britain. They do not, however, prevent people from dredging the bottom of the channel to clear it of sediment that washes down the Tamar and Plym over the year. The main channel is dredged as are the MoD harbours as well as some smaller harbours and marinas and each year a total of around 72,500 tonnes of sediment (eroded soil) is gathered and shipped a few kilometres down the coast to be discharged into the sea once more. Although there are restrictions on where can be dredged and where the material can be deposited there is never going to be a designation strong enough to prevent it from occurring because Plymouth Sound is too valuable to us. While sediments continue to wash down the river, freed from the farm land on either side, they will keep smothering marine life in the Sound and clogging the routines of people who will continue to ship our fertility from the harbours further out to sea and no conservation designation is likely to stand in their way.

As I realised that until our needs could be met from the land the seas would be impacted regardless, I started to turn my attention up stream, to agriculture.

Permaculture grabbed my attention because it focused on the intersection between people and nature. Its founders,

David Holmgren and Bill Mollison, took a logical approach to the task of redesigning farming and started by assessing the differences between 'wild' and human created systems. Holmgren and Mollison gathered up examples of different farming systems from around the world and compared them to natural systems. They came up with a design system that used ecology and traditional agricultural practises to guide modern food production and called it Permaculture, from a contraction of permanent and agriculture.

Some of Holmgren and Mollison's insights included that nature stacked many layers of plants into systems, natural ecosystems were highly diverse, everything in wild systems carried out more than one role and that native ecosystems were highly connected. These traits allowed wild systems to be incredibly productive whilst requiring very few inputs and producing almost no waste. Traits that they sought to replicate within farmed systems.

Permaculture lead me to a whole world of alternatives to dive into and explore; Keyline Design, Holistic Management, Agroecology, Masanobu Fukuoka's ideas around Natural Farming and Mark Shepherd's around Restoration Agriculture. They all got closer and closer to a central theme; how to mimic nature in our own farmed ecosystems to ensure their long term productivity, stability and resilience. In other words, how to harness our ability as an ecosystem engineer to create systems that benefit life as a whole.

Chapter 5

Regenerating Agriculture

Agriculture is an expression of our ecological roles as a keystone omnivore, an ecosystem engineer, a hyperkeystone species and a keystone mimic. It is the landscape that we co-create to sustain ourselves and it has the capacity to increase diversity, raise stability and offer resilience when interwoven with the wild. In this chapter we will dive into the world of regenerative agriculture and explore some of the practises that have sprung up around the globe that help to heal landscapes and merge our lives with those of the wild once more. Before we immerse ourselves in restorative food systems, however, I want to address the elephant in the room; agriculture needs to feed a lot of us reliably.

Feeding the Ten Billion

The global population level is set to rise to just shy of 10 billion by 2050 and there are many articles and newspaper headlines warning of our inability to cater for that many people adequately. It is worth digging into the numbers behind these headlines, however, because, in raw calorie form, we *already* produce enough food to feed this number of people and some spare to boot.

For the 2018/19 global harvest of cereals, assuming a population of 10 billion, there are approximately 660 kcal of wheat, about 956 kcal of maize, around 500 kcal of rice and a small quantity of barley, approximately 126 kcal, per person per day. This makes a grand total of 2242 kcal per person per day when distributed between 10 billion people from the top 4 cereals alone. The current global population is only 7.7 billion and if everyone alive today were fed from these cereals there would still be enough spare to feed roughly the population of the USA and Europe again, twice.

With these numbers it is staggering to think that anyone in the world goes hungry, however, around 30% of the global calories produced are currently wasted and around 33% of the land area used to grow cereals is harvested to feed livestock.

On top of our global cereal yield we also have all of the calories grown in orchards or raised in vegetable beds, those contributed by the smaller cereal crops such as sorghum, millet, oats or rye, and the calories contributed by grazing livestock. In order to feed the world adequately we don't need to produce more food, we just have to change our food system so that what is produced is delivered to where it is needed and people around the world have sufficient money to buy it or land to grow it.

Although we are obviously capable of feeding everyone with the agricultural system that we have now it has some downsides. Agribusiness contributes substantial sums of greenhouse gases to the environment, it degrades both freshwater and marine ecosystems with its run off of fertilisers, pesticides and eroded soil, it fails to support a vast array of plants and animals previously considered farmland specialists and it does not distribute food fairly with obesity and malnutrition both common across the world.

It is fairly obvious that the global food system as it stands now is broken but opinions differ as to how best to fix it and, indeed, what fixed would look like. There are generally two camps that people fall into, the land sparers and the land sharers.

The land sparers feel that wilderness needs to be segregated from people in order to protect its wildlife and propose allocating sections of the world to human development and leaving other areas, including the predominantly tropical biodiversity hotspots, to nature. Such a proposition would support highly specialised wildlife including those less adaptable to human interference including many apex predators. By ridding areas of the earth of man it is hoped that the earth system could function sufficiently well from the areas spared to repair itself.

This strategy relies on all of our needs being met indefinitely from the places cordoned off for our use and often assumes fairly intensive production techniques to achieve this. Industrial agriculture currently has no way to renew continually cropped soils indefinitely nor any way of capturing fertilizers, pesticides, herbicides and climate altering gasses that leak from such systems. Land sparing is further criticised from a social standpoint. The biodiversity hotspots that are removed from human habitation under a land sparing model tend to be in less developed regions of the globe or under indigenous management. To many, including the people who live there, land sparing appears very much like another land grab from developed nations penalising both indigenous groups and small scale subsistence farmers. There are also legitimate queries surrounding the wisdom of concentrating food production in the hands of a few very powerful companies in restricted areas around the world. If anything goes wrong with distribution networks or a capitalist system fails to distribute food evenly due to the unequal purchasing power of different communities people could be cut off from food sources altogether.

Land sparing requests that we downgrade our interaction with our food to a simple monetary transfer as we have done with our wild lands and entrenches the view that humans are somehow apart from the natural world and that land cannot be wild and peopled at the same time.

I also feel that as people try to conserve our wildlife by freeing it from people they potentially open another door to its demise. Without connection to nature we tend to forget our dependence upon the rest of life. Without daily ramblings in natural settings, without pulling fresh carrots, picking the first strawberry of the year or gutting a goose for Christmas people drift ever further from an understanding of what nature is, and what life is. They lose their understanding of it, their empathy for it and ultimately the deep love of it and the intuitive understanding that comes from the knowing that it is indivisible from ourselves. When Henry David Thoreau yeaned to 'know an entire earth' he not only expressed a desire for an *entire earth* but also a desire to *know it*. Not glimpse it from a viewing platform, purchase it in a shop or take a guided tour across it for a week, but to *know* it. To be a part of it. If both food and the wild are set beyond the grasp of people they become fictional ghosts of childhood days, relics that can be sacrificed in the harsh light of day. And then how long will 'strictly protected' areas remain protected, when we forget in the generations that come why they were protected in the first place and our descendents realise that money lurks below their feet or in their timber.

To me the only hope for restoration of the earth and of our own health is through integration. By entering into conscious ecological relationships with the world and co-creating something regenerative for the future.

This viewpoint can be categorised as a land sharing one and is the other suggestion of how we fix our food system. Land sharing has some substantial social benefits over land sparing with food production dissipated across the world and in the hands of local, small scale farmers. Such systems give local populations autonomy and food security and often produce a far more culturally appropriate and nutrient dense diet than foods coming from agribusiness systems. Small scale, localised production also preserves cultural diversity and indigenous knowledge as well as a wealth of local breeds, varieties and species suited to local areas and production systems.

Land sharing systems are often criticised, however, because they provide very little wilderness devoid of people where species such as apex predators can dwell and over worries surrounding how much food regenerative systems can actually produce.

The figures quoted above for global cereal production of the big four grains come from mainly non-organic systems and it is true that organic grain production can nowhere near rival chemical based grain production. From this starting point it is rapidly assumed that small scale, organic food production systems in general cannot contend with chemical based agribusiness. This is, however, not a fair generalisation to make.

Modern high yielding grain has been developed alongside a suite of chemicals to support its growth for the last few hundred years and the very recent research and funding that has been directed into organic grain has been vanishingly small (less than 1%). Moreover, grain is a difficult thing to produce from a regenerative farming system in the first place even if equal research effort had been put into varieties suited to regenerative and chemical systems in the first place. The framing of the question is far from fair.

If we broaden the question and compare conventional grain yields with the calories and nutrition produced by the same acreage of diversified, ecological cropping systems the playing field levels somewhat. It is still nowhere near a fair comparison because of the lack of research and breeding of regenerative crops or large scale, fully supported trials of them but there are some places that we can look when attempting to gauge whether such alternative food production systems could feed 10 billion of us. The first evidence that we can gather comes from the practise of intercropping.

Farmers are finding, almost without exception, that more food can be produced from complex mixtures of plants than could be produced on the same acreage if the crops were grown separately. A land equivalent ratio (LER) is one way of describing this effect. It compares the yield from a monoculture stand of a crop with the yield from the intercropped stand per unit area. A result of 1 means that the yields are the same with results of less than 1 meaning that there is a reduction in yield and results of more than 1 that there is an increase in yield. Some intercrops that have been experimentally tested include peas grown either with garlic or turnips, wheat grown with broad beans and maize intercropped with potatoes all of which yield positive results (results above 1). Experiments have also been done on agroforestry systems, systems that include trees, including tree strips grown amongst cereals or vegetables, with similar positive results.

Interestingly a similar effect occurs when different species of livestock are grazed together. It is common to work out how much livestock an area can support in livestock units (LU). One livestock unit is equivalent to one dairy cow (in milk and weighing around 500kg) and other livestock units are given in relation to that of the dairy cow, for example, a beef cow is 0.75 LU, ewes are considered to be between 0.06 and 0.11 depending on breed and sows about 0.44. Different types of land also have different ratings as it were, for example, a lowland field may be expected to carry 3 LUs per hectare where as a heather moorland may carry no more than 0.1 LUs per hectare. Seemingly regardless of the land type more LUs can be carried if the units are of different types of stock. An old farmer's saying which seems to bear some weight is that for every cow calf pair stocked three sheep and lambs can be carried 'for free'. This is because they are using different portions of the forage on offer, complementing one another's grazing styles. This starts to go some way towards highlighting the potential benefits of small scale, integrated and diverse systems.

Perhaps the best trial of organic production over a large area has taken place in the Indian state of Sikkim. Gradually, over the course of ten years, Sikkim became the first organic state in the world. As expected yields initially fell as the ecosystem, deprived of the chemicals on which it had been running, were cut off. Gradually, however, the ecosystems started to repair themselves. Soil stared to build in fertility and pests and predators fell back into steady cycles with one another. Three years after turning organic yields were as high as the chemical supported systems yields had been. This hints at the acreage not having to be vastly raised to produce comparable quantities of calories and nutrition and means that, even in a land sharing model, we can probably still afford to leave large areas minimally peopled.

At the end of the day one small point is impossible to deny. As Dr Miguel Altieri puts it:

There are around 1.5 billion hectares of farmland, and 80% of that is in the hands of just 20% of mostly large-scale farmers. Those farmers only produce around 30% of food globally leaving a huge ecological footprint, whereas the remaining 80% of farmers with only 20% of land and using less than 30% of the water and energy are producing 50-75% of the food.

It is impossible to know how the future will play out. We know that the system as it stands is not working and that it is contributing to problems such as climate change and ecological collapse. We have ample reason to doubt that agribusiness and ecotourism, seated within a capitalist system, can provide nutritious and dependable food in a culturally appropriate way indefinitely from delineated areas of land whilst screening toxins leached out into the wider world.

We will also never know, until we have tried it, whether people can co-create regenerative systems that can feed us nor how much land that would take up across the globe.

Regenerative agriculture is likely a good option, however, for feeding 10 billion people *and* securing wild lands where the wild incarnate, as Aldo Leopold called them, species can thrive.

Ultimately, we know that agribusiness depletes soils

and degrades ecosystem processes and we also know that regenerative agriculture builds soils and restores ecosystem processes. Even if now yields are comparable in ten years time I know which system I would prefer to be fed by.

A Regenerative Mess

One of the main reasons why farmed systems harbour fewer species than wild ones (other than because of the application of agrichemicals) is that they are simplified. They are landscapes of disturbance that have been turned inside out with blanket disturbances applied across them. Neat fields that are mown and reseeded with tightly clipped hedges or orderly fences running between them punctuated perhaps by angular plantations of even aged trees in rows. This is how we manage our landscapes; as simplified and delineated blocks exposed to regular and uniform disturbance.

Modern regenerative approaches can be seen as reversing those trends, moving towards a new form of productive, integrated complexity populated by trees and permanent ground covers to heal the soil and capture sunlight, imbedded with water ways to mend water cycles and supporting a wide range of wild and domesticated species in diverse food webs. Regenerative agriculture builds the three key properties of healthy and mature ecosystems; productivity, stability and resilience. Although it does not claim the title itself, it can be seen as a landscape presided over by ecosystem engineers allied with the decedents of wild keystones, our livestock.

Water and Earth

Rivers always cut the path of least resistance through a landscape. They hew their course from the softer rocks following in the footsteps of their frozen glacier ancestors as they wend their sinuous path to the sea. As a river turns a bend its full force is thrown against the far bank, cutting away at the land to create the exposed faces so loved by sand martins and king fisher's alike. As the river turns it also leaves behind it a smooth beach and area of still water, home to ground nesting birds, invertebrates and reptiles. As water weaves from the hills to the sea, from chattering streams of mayfly and trout to sluggish and silty old age meanderings, people corral and hurry it. Its ever rasping jaws are not welcomed on the outside of a bend nor its precious collections of neatly sorted gravels, sands and silts atop a road or ford. Above all its periodic swelling and unstoppable seep across floodplains, now home to cities, is forbidden.

For centuries farmers have tried to speed water up as they have hurried it through drainage channels, along canalised and straightened rivers and out to sea. Heavy machinery and annual crops both require well drained soils and the shedding of water is a major objective of British farming.

Shedding water rapidly, however, has a few side effects. After heavy rainfall the water runs rapidly into drains and ditches that guide it quickly towards rivers slamming into waterways all at once and increasing the likelihood and severity of flooding. Fast gullies and channels of water erode soil more easily and carry it further as they hurry along, depositing thousands of tonnes of soil into the sea each year. The eroded sediments also carry nutrients out of our soils and into water bodies where they damage aquatic life.

The hurrying of water from the land also means that when it ceases to rain the land dries out very quickly and farmers often turn either to the mains or the network of rivers and aquifers to supply their stock or water their crops.

Alternative agriculture tends to approach the matter from the opposite end, it takes a leaf out of the beaver's book and seeks not to speed water up but to slow it down. Regenerative farmers often put the bends back into streams and establish rough mats of vegetation to soak up and quiet water flows. They allow it to settle into soils where it will be held until needed. Regenerative agriculture starts to meld land and water once more.

One man who has done perhaps more than any other to redirect water from being a problem into an asset is the Australian P. A. Yeomans. The idea of slowing water down to strip it of its erosive power and of spreading it out so that it might infiltrate into the soil was well established but Yeoman's created a system whereby he took this one stage further. He would drag a long tined plough through the soil just off the contour parallel to what he called the Keyline. The plough sliced through the sward and created a small subterranean tunnel in the soil. He created a landscape below which ran parallel tunnels that led water from areas where it naturally collected, in the valleys, gently back up onto the ridges, from which it naturally drained. In dry and dusty Australia, where a lack of water often prevents organic matter from decomposing, Yeomans was able to create rich soil and re-establish vegetation very rapidly. The new vegetation intercepted the water as it fell, reducing its power, and its roots helped to guide the water into soils reinforcing the change.

Yeomans also built a series of larger ditches that carried their precious cargo of water across the hills to deposit it in carefully positioned dams high up in the landscape. These dams could then feed all of the land below them allowing him to provide irrigation to his stock and crops without needing to draw water from natural sources. His farm acted like a sponge, pulling in water during rainy periods and releasing it during dry periods, to buffer against downstream flooding.

By weaving water through our landscapes farmers can moderate increasingly erratic weather patterns, build connection between wet and dry and embrace our ecosystem engineering role of creating areas of difference within monotony. Water creates a different type of habitat that provides food for a rich diversity of wildlife and can also provide another source of food for us.

Fish are perhaps the most common aquatic animal eaten and they were one of the first stops that I made when entering the world of alternative agriculture. As a child I was often to be found poking about in the pond at the bottom of the garden and as I grew, I graduated onto lakes, reservoirs and rivers. I was fascinated by the diversity of life that lurked below the water surface and I hauled out net after net of plants, invertebrates, fish and sludge. The more interesting finds I would tip into a jam jar for closer inspection and over the years many made their way back into tanks in my bedroom. I collected frogspawn and toadspawn and watched as the tiny jelly dots grew tails and then legs. I gathered the nymphs of dragonflies and damselflies and peered at them as they grew with each shedding of their old skin until, eventually, they emerged as winged adults. I kept beetles, bugs, snails, leeches, water fleas and hydra. And, of course, fish. I had tanks replicating the fast streams in which bullheads live and others the leafy green ponds home to sticklebacks and minnows and also the obligatory goldfish, although mine were in a tank 6ft long.

By the time I encountered regenerative agriculture, fish keeping felt like very safe ground and, after a brief scan of the information available, I decided to construct an aquaponics system. When fish are reared in normal aquaculture systems more fish are stocked than would be present in a natural water body of the same size. As a result these fish eat more and produce more waste which needs to be stripped out of the water to ensure that the fish remain healthy. Only bacteria have the ability to convert the waste ammonia produced by the fish into less toxic molecules tolerated by them and so the levels of bacteria that a system can support become critical. Bacteria like to live on solid objects and in a wild lake they blanket the thousands of plant stems, leaves and twigs which provide ample room for the numbers of bacteria needed to process the wastes of the lake dwelling fish. When fish are stocked at greater densities, however, a filter is needed which provides the vast surface area required for billions of bacteria. Aquaponics has turned the problem of providing sufficient surface area for bacteria into a benefit by filtering the water through grow beds containing a vast quantity of plant roots. The grow bed media and plant roots combined provide an astonishing level of surface area over which bacteria can convert waste ammonia into nitrate. Nitrate is plant food and the plant roots absorb it to grow food for human consumption.

I established a 3ft tank connected via a hosepipe to a propagator on my bookshelf that functioned as a makeshift grow bed. I knew that I wanted to produce some fresh salad in the grow bed but what to put in the tank?

The only real input into an aquaponic system is fish food and so which species of fish to keep, and what food they like to eat, becomes important in determining the environmental impact of the system as a whole. In warmer regions the bacteria that live in the digestive systems of grazers also dwell in the digestive tracts of fish allowing them to break down and access nutrients from plant foods, however, in Britain our fish rely on animal protein, lacking the microbial partners associated with their warm water relatives.

My first thought, when considering edible freshwater fish, was trout. Trout, however, along with their relative's salmon and grayling, live in cool and fast flowing waters feeding almost exclusively on small invertebrates as they drift downstream or blunder onto the water surface. These fish need pure water and a diet high in animal protein and I could not see how I could cater to their tastes without relying on purchased feed (which is often composed of wild caught fish meal).

I turned my attention instead to our less well known edible fish species. More commonly encountered in lakes and the slower reaches of rivers these fish can tolerate poorer water conditions and tend to rely on invertebrate-plant medleys making them far easier to cater for. They include the herds of bream that patrol slow waters grazing on bottom sediments and the flocks of tench that venture out to join them from the shelter of gullies, stirring up debris to re-fertilize the water column as they go. Around these 'large grazers' flit a variety of smaller fish including the fast breeding and unfussy roach, the shoals of smaller rudd and the tiny minnows all of which make easy prey for the carnivores. Packs of small perch dart about catching anything from invertebrates to roach and, as they age, even smaller perch. Pike are our other aquatic apex predator; they hang motionless in still weeds waiting to strike at anything that will fit into their mouths including water birds and unwary rodents. I once saw two pike, both dead, which had become locked together as one had attempted to swallow the other head first, only to find out that it was two big. Pike too will prey on smaller pike, just as perch do, regulating their own population sizes.

All of these species can, and have in the past, been eaten. In America today yellow perch, a close relative of our perch, is eaten and attempts are being made here to rear our own perch, although somewhat hindered by its tendencies towards cannibalism in single species tanks. Tench too are considered fine food fish in other areas of the world and are very easy to rear having a thick mucus coat that protects them well from almost all diseases and parasites. Although some of these species were intriguing I chose to start out with the far more readily available carp.

Common carp are native to Asia but possibly reached the British Isles with the Romans and were grown in stew ponds by monks for generations. They are hardy fish and very unfussy eaters and seemed the perfect place to start. Carp are omnivorous and are quite happy to try most things including many kitchen scraps like bread, sweet corn and scraps of meat.

People are also feeding them on pre-fermented plant matter and culturing invertebrates for them including black solider fly larvae and house fly maggots raised on waste.

With my carp installed my micro-aquaponics system was ready and in it I grew a wide array of salads year round. Over the years, as I became more adept and my carp got bigger, we outgrew my propagator and moved into a series of IBC tanks in a small polytunnel. And then into a larger tunnel with a pond that took me 5 days to dig, replete with its own off grid electric system.

Amphibious agriculture in Britain is still very much in its infancy and under researched but it shows much promise. People are using wetlands as grazing areas for livestock, they are rotating ducks around ponds that are then drained and used for vegetable production and they are floating pontoons of salads on fertilized pools.

Wetlands are also perfect processing factories for dirty water as Jay Abrahams is demonstrating with his WET systems, convoluted wetlands that can transform sewage into water clean enough to bathe in.

The possibilities of working with water are almost limitless and largely untapped. We have focused such attention on draining our wetlands away that we have largely forgotten what rich resources they could offer. Regenerative farmers are only now exploring the tip of the iceberg and starting to reverse the divide in our mindsets between productive land and damaging water.

Miraculous Soil

Weathered from rock, bound together by life and holding within it both air and water soil unites all of the elements of the biosphere and underpins almost all food production.

The first soils were eroded from the bare rocks of the earth and accumulated in nooks and crannies eventually being joined by roots as the first plants crept from the sea. As plants evolved, their roots burrowed down into sediments stabilizing them and stocking them with food. Microbes wormed their way into soils and over roots, stems and shoots gathering resources from the plants and protecting their hosts from anything that might damage them in return.

Plant roots excrete a wide variety of substances including sloughed off root caps, acids, sugars and other metabolites. In fact up to 70% of the energy that a plant fixes during photosynthesis can be exuded from its roots to feed soil life. Plants can even tailor what exudates they release to attract certain types of microbe that can perform certain functions for them such as acquiring a nutrient which they currently lack or warding off a disease that threatens them. Some microbes are capable of some truly amazing things in defence of plants such as the bacteria *Streptomyces lydicus*, which can produce a range of chemicals that combat many fungal plant pathogens and *Bacillus pumilis,* which intercepts pathogenic fungi as they land. Plants also have a version of an immune system (called system acquired resistance) that microbes help to run, as they do our own. Some strains of *B. pumilis* seem to be able to induce the plant's defence system to recognize certain pathogens allowing the plant to respond, sometimes before the disease has even arrived yet! Needless to say plants take good care of their microbe allies.

Microorganisms flock to capitalize upon the plant buffet on offer and their populations can be up to one hundred times higher around plant roots than in open soil. Predatory microbes, including protozoa and nematodes, feed on a plethora of smaller microbes digesting and releasing the nutrients held within their miniscule bodies. These predators keep their communities dynamic by cycling nutrients between plant, herbivore and carnivore just as their larger counterparts in the world above do.

Plant litter rained from above is the other main source of food for microbe populations and the organisms that take part in this food web are no less diverse or dynamic than those feeding off root exudates. When a sudden input of matter is delivered onto the soil surface a healthy soil microbe population will respond by rapidly multiplying to take up the excess nutrients. Fungi tend to be some of the first, fast acting sugar seekers on the scene, joined by some bacteria, invading freshly fallen litter to start the breakdown process. Hunting the fungi and bacteria are protozoa, the miniature wolves that liberate stored nutrients from its biological packaging. The protozoa provide a drip feed system to the plants by freeing the stored nutrients from the bacteria's bodies as they steadily consume them.

Certain species of fungi bridge the gap between plants and soils by sheathing plant roots or invading plant tissues and also stretching out through soils, linking the two together. These mats of mycorrhizal fungi as they are called extend across landscapes, merging with one another and multiple plant partners, binding soils and plants in one dynamic web. The fungi exchange nutrients between themselves and multiple plants meaning that healthy plants in prominent positions can be linked by underground fungi to plants in less optimal areas. Information is also shared on this subterranean network as plants suffering from pests or pathogens warn their neighbours of the approaching threat through the fungal web. Essentially the mycorrhiza distribute energy, nutrients, water and information around an ecosystem supporting the plants that need it most and uniting a whole landscape into one coordinated fusion of plant, microbe and soil.

Springtails, fly larvae, woodlice, bristletails and silverfish continue the breakdown of leaf litter gnawing holes in debris creating a higher surface area over which bacteria and fungi can bloom. There are also a wide array of mite species involved in decomposition with the smaller oribatid mites eating fungi and detritus and the larger mites eating the smaller ones. Rotifers eat the lot, leaf and microbes combined, shredding the material still further. The surface area is increased at every stage and the mixture is added to by millions of defecating life forms altering its structure and nutrient content the whole time.

Centipedes and spiders prowl through this jungle in miniature catching and eating most things smaller than themselves and slugs and snails are like the megafaunal grazers, gliding through the mass of matter eating plant debris and being preyed upon by the fast moving ground beetles and birds.

The soil is a dynamic food web in its own right comprised of grazers, browsers, shredders and carnivores. It also has its own keystones and engineers. In Britain we have 27 species of earthworms some of which live and feed in decaying matter, others that burrow into the surface layers of the soil and a couple that form deep, vertical tunnels that they maintain for up to 20 years. The vertical tunnelers have an enormous impact on soil by excavating down to 5m or more and lining their burrows with decomposing leaves. These worms can reach population densities of 20 to 40 tunnels per square metre, massively increasing how much water can percolate into a soil and how well plant roots can penetrate. As worms process soil they also stick it into stable aggregates, add plant growth stimulants and alter soil pH. Worms are our soil sculptors and, just like us and beavers, have the duel titles of keystone species and ecosystem engineers.

The array of interactions between plants, microbes, soil and soil life binds it into a single entity. They are one and the same and they create one another. Plants do not grow in a soil, they form it.

Eventually humus, the wonder that binds our soils together, is all that remains of the discarded plant matter, the soil's carbon, water and nutrient store that underpins an ecosystems long term health. It is forged from a healthy relationship between soil life and abundant plant matter and in a healthy soil it is constantly being formed, transmuted and broken down once more. The rate at which farming practises build or use up humus is key when thinking about climate change as well as providing sufficient nutrients and water for growing crops. Conventional farming has a fairly bad track record when it comes to soil humus content primarily because of its addiction to tillage, burning through humus at an alarming rate and adding little back in return.

A Disturbed Sleep

As we discussed above it's not disturbance per se that is negative, it is the scale on which it is practised and this holds true for soil. Small areas of disturbed soil are an essential part of maintaining a dynamic mosaic of vegetation types and many invertebrates require holes in the vegetation carpet in which to complete some aspect or another of their lifecycle. Modern farmers, however, approach the task in a much larger, more frequent and more uniform way, cleaving plants from soils for extended periods. Before we go on to look at some possible ways of integrating beneficial levels of soil disturbance into our ecosystems lets dive in and explore some of the issues that arise when the relationship between plants and soil breaks down.

Tillage has been the backbone of Eurasian style agriculture since its birth 10,000 years ago and over this time its shortcomings have been brutally exposed. Research into rates of soil erosion has been carried out around the world with Sheffield University scientists concluding that UK soils have about 100 harvests left in them and the UN that the world's soils have about 60. On top of soil erosion the flood of oxygen into soils combined with the growing of greedy crops that contribute very little organic matter lowers humus and soil carbon levels year on year. This is bad news for the farmer, because without some intervention yields will fall, but it is also bad news for the climate with the carbon locked up in soils released to the atmosphere. The bare, dark coloured soil also tends to heat up more than the surrounding plant clad land retaining heat and evaporating water from the soil surface. Water vapour is also a greenhouse gas and has, according to soil ecologist Christine Jones, proliferated in the atmosphere since the industrial revolution due to tillage.

So why, with so many negative side effects, do we still till large areas of farmland?

The central problem is that all of our main crops, the cereals, are annuals that need to be sown into freshly disturbed soil each year. This requires that those lands are ploughed, away from the sediment depositing rivers of their home. In organic systems tillage is also used to suppress weeds and to bring fallowed land, still used for fertility building and pest suppression, back into production.

One of the central questions of regenerative agriculture is how to grow enough food from soils that retain an intact plant cover year round, fostering the relationships between plants, soil and soil life.

Around the world a plethora of alternatives have been suggested. One strand of thought is to breed perennial versions of our cereal crops. This at first sounds fairly straight forward; most of our cereal crops were bred from perennial ancestors and so back breeding our modern varieties with their longer lived relatives would be expected to bear fruit. Unfortunately this is often fraught with problems as the ancestors of our cereals are generally not well known and the domesticated versions have come a long way from their wild roots. The hybrids created so far don't tend to retain the reliable, super high yielding qualities that modern grains have been bred for. In hindsight this might have been expected. Annual cereals are eager to produce a heavy head of grain because for them it is their one chance to reproduce and they will die at the end of the year regardless of whether they fruit or not. Perennials arrange their life in a different way, they invest in producing a strong individual plant that can survive for many years and if they do not reproduce in a poor year it is not the end of the world for them. It is difficult therefore to persuade these perennial versions to yield as highly and reliably as their annual counterparts.

Another line of thought has lead people to attempt to grow our traditional annual cereals *without* ploughing. The 'no-till' farmers as they are called generally sow an area to a lower growing crop which carpets the ground. The ground cover crop is then killed off before the cereal is seeded directly into the resulting mulch of plant matter. This approach is also not quite as straight forward as it sounds, however. There are very few suitable ground cover crops that will grow up over the times of year when they are needed to shield the soil surface and, more importantly, vacate that position as the crop plant replaces them. The only sure fire way of killing off the ground cover is with a herbicide which obviously comes with an array of its own difficulties. Numerous groups, including the Rodale Institute, are working on selecting cover crops and organic methods of killing them off for different organic systems, however, it is far from a precise or guaranteed science.

On smaller scales, of course, the problems associated with weed suppression melt away. More or less any cover crop can be used to fit in between cereal crops because the problems associated with killing them reliably vanish at the human scale. Black plastic or another mulch can be used to starve them of light allowing crops to be planted through.

This was one of my first jobs when I took up a role as a grower on an organic market garden and it remains one of my favourites. There is nothing quite as satisfying as laying wet strips of cardboard over a weedy bed, burying them in a layer of compost and planting out the next crop. Very quickly the mess of weeds can be transformed into neat lines of vegetables, something that deeply satisfies my inner *sapiens* even if my inner ecologist shudders at the thought of transforming complexity into uniform rows.

Interestingly the problem of tillage also diminishes at the human scale. A small tilled plot amidst a network of established vegetation is far from disastrous. An area of useful difference amidst stabilising vegetation and the disturbed gap heals over quickly, a small nick out of an otherwise thriving ecosystem. The downsides of smaller patches of grain, however, are that less can be grown across the landscape as a whole and either many hands or adapted machinery are required to make it viable.

The difficulties of finding a broad acre solution to producing our staple carbohydrates have lead other people to search more widely and abandon cereals all together. These pioneers, including J Russell Smith, have looked to trees instead. One of the more promising substitutes for cereals is hazel, a hybrid of which has now been bred to rival the kcal produced per acre by chemically grown wheat. They are generally planted commercially in rows much like a modern orchard and can yield 3 to 4 tonnes of nuts per hectare; for the next 100 years or so. Sweet chestnuts are another option. Nutritionally similar to brown rice they live for even longer, over 1000 years in fact. They were brought to Britain from southern Europe by the Romans where they were historically grown across the hills of Europe while the cereals grew on the valley floors. Walnuts too, once the pride of Persian Royalty and carried to Britain along the Silk Road, may be making a comeback with new reliable varieties being bred to cope in climates such as ours. Martin Crawford has been spear heading British organic nut growing trialling many varieties of the more familiar nuts, especially chestnuts, as well as some less familiar ones including bladdernuts and heartnuts as well as hickories and almonds at his site in Devon. Tree crops can form deep and lasting relationships with the soil. Large trees reach down through many soil layers and collaborate with mycorrhizal fungi to act as redistribution mechanisms for water and nutrients as well as recognizing many pathogens and pests that have come and gone over the years. Trees have time to build truly vast and deep interactions with soil and can capture amazing quantities of light energy to be directed into both human food production and soil fertility. They do come accompanied by some problems, however; people can be unwilling to eat diets based on nuts and currently there are not enough established nut platts to feed us. We also lack the infrastructure to process nut crops on large scales.

On their own none of these solutions can hope to rival the conventional cereals, however, when combined they provide a formidable package. Smaller networks of tilled or no-dig cereals, perhaps interwoven with annual vegetables, can provide the open ground below and between nut trees. The nut trees in turn can provide a multitude of benefits to the system including enhancing water infiltration into soils, minimising soil erosion and protecting soils from harsh wind and heavy rain. Trees and annuals together offer us a good diet that can also replenish our soils but work best when combined with a third ingredient; grassland. Grassland can restore soil fertility, lower arable weed populations, fix nitrogen and capture escaping nutrients and soil.

Productive Grassland

The second step in my journey into agriculture was on a mixed organic livestock farm. I had returned from the 'year of obligatory travel' after university during which I had worked on a wide range of farms and volunteered on conservation reserves that used stock (rather than roundup) to manage the landscape. I had made contact with a local farmer who had just signed a tenancy agreement on a 100 acre farm and set off across the two intervening hills to meet him. I knew the public footpaths that crossed the land well but I had never ventured near the farm buildings before. I crossed a cattle grid onto the farm track and followed it as it twisted and turned through straggly birch and oak woodland towards the farm buildings crouching low on the hill. I could see the silhouettes of two people hauling feed sacks out of the back of a Land Rover and handballing them into a barn. I headed their way. After a brief introduction we set off to check the sheep and it wasn't long before I was thrown in way over my head. One of the manx loughtan ewes had run into difficulty lambing and we had to intervene, quickly. Corralling her into a makeshift pen, gently realigning her lamb (who had attempted to come out all legs and no head) and assisting with the rest of the birth was my first introduction to livestock farming. As I sat, crouched in the back of a truck with a wide eyed ewe and two, jet black lambs still trying to come to terms with the world, I wondered what I had let myself in for.

I quickly settled into the routine of the farm, it seemed second nature. Get up, feed and let out the poultry, milk the goats, check the sheep and cattle, feed the pigs and then home to the caravan for breakfast. Life was fast paced and hectic, there was always more to do than there were hands to do it or hours in the day and each new season brought its own challenges. I had arrived in spring and long days and nights were spent checking on ewes and lambs, bringing the expectant flock in overnight so that they could be watched more easily and leading them out over the day to graze. We took shifts to stay with the ewes over night and the gentle chewing, the occasional bleat of a lamb from the individual stalls on the near side of the barn and the glint of four hundred eyes through the amber gloom became a part of life.

Spring rolled into long summer days of anxious weather watching and then frantic hay making racing the rain to get the year's crop safely stowed before the heavens opened. My job was often to drive the telehandler, scooping up bales onto a trailer to escort them into the barn and stack them neatly ready for winter feeding. I spent long days bouncing over concrete hard ruts in the ground in a sweltering cab with the radio on full blast to drown out the sound of the whining engine. Summer rolled into autumn, the lambs were rounded up and sent off in batches to come back in white cardboard boxes, and then autumn drifted into winter. With Christmas drawing ever closer, and the turkeys and geese looking ever fatter, anxiety levels started to rise once more.

I was vegan when I went to the farm but after several months of catching, slaughtering, plucking and gutting chickens and ducks it all started to seem a bit ridiculous. I had began to drink the goat's milk, making porridge with it each morning and cheese from the surplus, and I took my first tentative bites of meat in the form of gizzard soon after. When preparing a bird for market it is plucked and gutted before the head, neck, feet and sometimes the end joint of the wings are removed. There are certain bits, the giblets, which are then bagged up and put back inside the carcass before the bird is trussed up and packaged. The giblets usually consist of the neck, liver and heart with all of the other innards being discarded. The gizzard is one of these, usually discarded, organs. It is muscular, slightly iridescent and somewhat reminiscent of a shellfish and, just like a shellfish can be cut open and its inner lining peeled away, a bit like peeling thick sunburn off shoulders, to leave a lump of meat. It is good pan fried on toast and it became my weekly mid day snack.

Two weeks before Christmas everyone was running flat out. The first year we killed the turkeys in one batch and allowed them to cool before plucking them. That was a big mistake. Feathers come out of a warm bird quite easily but once it has cooled down, not a chance. I remember standing in the door of the barn, the upside down and lifeless bodies of 50 Norfolk black and bronze turkeys rotating slowly in the breeze behind me, looking out over the white yard as thick flakes of snow slumped silently to earth. I had blisters on my hands and chilblaines on my feet. I also had the strongest sense of what Christmas was, and had been, and of what life was, than I had ever had before.

The years rolled on and I got better at lambing, faster at plucking and gutting and more adept at catching a guinea fowl in mid flight or a lamb mid leap. I also started to watch the world in a completely different way. I had developed a farmer's eye for the rate of grass growth and a feeling for the condition of stock. A delight in the uniform green of a field and the satisfying balance of grasses and clovers that would mean a good diet for the grazers. I was comfortable with machinery and power tools and used to the idea of 100 horsepower of tractor there when I wanted it. I had become accustomed to watching a grasshopper chirp merrily from a clump of tall grass before the head of a chicken swooped, like a gannet from on high, into the ocean of grass to gulp it down in one. I felt a level of satisfaction in knowing that the protein in the egg that the hen would lay tomorrow had come free from nature. I also started to feel a growing unease and a tussling in my brain as my two ways of seeing the world came into conflict; my inner conservationist had a very different appraisal of these agricultural landscapes.

The fields were too neat, too uniform and my eyes strayed to the tangle of brambles spilling from a hedgerow into the grass. There was life there, butterflies and birds, but there was also danger. It was only a matter of time before a lamb in full fleece would push into those brambles in search of an autumn snack and become stuck. I started to ponder too the impact that a field of greedy chickens had on the invertebrate population as the hens funnelled anything that moved down their necks. The need to keep the chickens safe from foxes, and also to separate them from the turkeys because of a potentially fatal disease that chickens carry and to which turkeys have little resistance called black head, meant that they were housed in a large, 6ft tall mesh pen that stretched up one side of a field. Initially the whole field had a smattering of docks but after the first summer a stark line had emerged between and the inside and outside of the chicken run. The docks outside had become lacy and thin, their leaves picked to the bone by thousands of shiny green metallic dock beetles which fell from the plants like rain when knocked. Inside the pen there was not a beetle to be seen and the docks grew tall and glossy, resplendent under the protection of their chicken patrol.

By artificially maintaining a high density of stock in an ecosystem were we displacing wild animals to find their food elsewhere? Were we, over the long term, running the farm into the ground by over exploiting the natural foods that the system offered? My views on the matter were to oscillate back and forth for many years and the battle ground was nearly always 'pasture'.

Grasslands fix nitrogen, mop up fleeing nutrients, capture energy from sunlight and create soil but they produce almost no food for humans without their partners; the grazers. The starting point for alternative livestock models is often that grazers and native vegetation support one another to exist and whilst this might sound fairly obvious within the context of this book it is actually a radical statement for a farmer as we shall see.

It is probable that in an unpeopled Britain patrolled by wolves the numbers of wild grazers would have been insufficient to dent the profusion of growth over summer, possibly excluding some close cropped grazing lawns. Over winter, however, the animals would more than likely have grazed the land hard back removing the course mats of grass that would have sprung up and slumped over during summer. The stripping back of old growth over winter would have prepared the land for next spring's flush of growth, allowing light to penetrate to the delicate growing tips as they pushed their way out of the soil and into the air. Incidentally this is still the role of many cattle and sheep employed by conservationists in their grazing programmes around Britain today.

With the Roman introduction of the scythe and the start of haying, stock could be kept at higher densities on smaller areas of land over summer whilst other areas were saved to grow hay for storing over the winter. The hay would have been cut at the end of the summer and the stock brought indoors to eat it from the comfort of a warm stall. In effect the grazing lawns were expanded into pastures, the abundance of summer growth was secured as hay and the stock that remained out over winter would have joined the wild animals in clearing up the remains of summer's abundance. The interactions had become more 'formalized' and the relationship between grassland and grazer had started to pull apart as we became the mediator between the two.

For centuries farmers watched their pastures grow and noticed that some plants grew faster than others. They then, very logically, tried to grow more of the faster growing grasses with the idea of being able to feed more stock and gather more hay. As grass seed became available over the 1600s and 1700s farmers started to sow a more limited array of very productive grasses. Today the various ryegrass varieties are almost the only ones grown having been bred for vast yields of lush leaves high in sugar. These fields of fast growing grasses can use an almost limitless quantity of fertilizer and convert it quickly into grass that can support a vast biomass of grazers. As livestock feed they target the tastiest morsels, historically picking out a nutrient rich and balanced diet from the wide array of plants on offer and, evolutionarily speaking, this has been a tactic that has served them well. In a modern world of ryegrass monocultures, however, their focused munching of high sugar grass allows other species to sneak back in. These interlopers, usually docks, nettles and thistles, are far less palatable than ryegrass and thrive, soaking up the nutrients on offer as the stock remove the grass from around them. The farmers notice the profusion of 'weeds' across their fields and wage war on them trying to keep them out, endlessly fighting against diversity for maximum productivity. Although a lot of animal biomass can be sustained on these simplified fields the mutually beneficial relationship between grazer and grassland has broken down. Grass, in farmer's eyes, has become the thing that farmers grow to feed to stock, not something that emerges from the relationship between animal and land.

Regenerative agriculture has sought to recreate these lost relationships between grazers and landscapes and rediscover the wealth of species that once grew in our pastures and meadows contributing to diverse, steady and stable food webs.

The person who has done the most to progress the idea of reuniting grazers and plants is perhaps Allan Savory, the mind behind Holistic Planned Grazing (HPG) as a part of Holistic Management (HM). The Zimbabwean ecologist existed in an area in which the African vegetation was dying. Researchers assumed that it was being overgrazed and so culled the herbivores in an attempt to allow the vegetation a chance to regenerate away from their hungry mouths, but the opposite happened and it continued to shrink. Savory watched the vast herds of ungulates maraud across the African landscape, hotly pursued and corralled by an endless array of predators, and a counterintuitive idea began to emerge in his mind. The grazers consumed everything before them and left in their wake only the remains of crushed stems and leaves, dung and urine; a hotbed of decay. Savory realised that in hot and dry environments, such as large areas of Africa, decomposition needed to occur within the moist and hospitable rumen of the grazing herds. It was only in their guts that fertility could be liberated from dead plant matter to become available once more for plants to take up. Far from destroying the vegetation it was the wild mega herds that had created it in the first place.

Based on his observations, Savory formed a system of grassland management in which he tightly corralled cattle within electric fences encouraging them to severely impact the 'grazing cell' before moving them on to the next one. The grass was allowed to grow up until almost mature, offering the highest biomass of most palatable grass, before the munching mouths were allowed to return once more.

Grasslands will respond very well to such harsh treatment having evolved alongside grazers. Over half of a grasslands biomass is below ground and as the top half is defoliated the bottom half sheds its roots proportionally leaving them in the surface layers of the soil to rot down. This is the primary mechanism by which grasslands build fertility and lock carbon away in soils and has sparked massive interest in the world's grasslands for their role in climate change mitigation.

Holistic Planned Grazing has yet to be tested experimentally, a job that science in its usual form is not particularly well suited to because of the inherent complexity of the system, however, a few people have made some fairly substantial claims. Savory, among others, has focused on the role that HPG can play in halting climate change jumping the gun slightly on the scientific backing and leading to criticism. What impact HPG across the world could have on the climate is probably unknowable but the results of its implementation across Africa and other drylands are quite staggering. Savory's ranch glows green in a sea of brown, a sharp halt to the desertification that creeps in the wake of disturbed grazers, but its potential impact to halt climate change is perhaps over stated. Only time, and a lot more research, will tell.

A cool wind grabbed at the rough edge of rock as I stood there, the first light of day touching the tops of the trees and trickling down a lazy bend of river on the plains far below. I watched the long shadow of a giraffe make its slow way across the valley floor towards the water and thought of the elephants, lions, leopards and wildebeest that I knew were down there too. I had seen them, watched them as they lazed about, something so familiar and yet so jarringly odd about them. But how did this relate to a land that I knew so well, to Britain that was both so similar and so different. As I stood high on an African cliff, with the sounds of day creeping up behind me, a swallow whizzed though the air and I wondered if it could be one of my swallows that spends its summer shitting in my hay barn. Would it see a difference between here, its winter haunts, and my rough pastures? Here it caught flies above rhino and between the legs of elephants; there it would catch them above cattle and horses. Was there a difference between this land and mine?

Savory's model has spread from Africa out across Australia and America and it has also arrived here in Britain. Many herald it as a new age of livestock management, one based on observing what happens in natural grasslands and mimicking them to best support our wildlife, produce nutrient dense meat and lock carbon away in soils. Savory's method was developed by looking at how grazers and grasslands create each other in his home country and mimicking that within agriculture. Does the same model apply as well within Britain, however, or did Britain's grazers have a different relationship to the land? The clues, once again, lie in our islands past.

Some claim that Britain once had herds comparable to those of Africa that were, just like Africa's herds, corralled tightly by predators as they migrated across the country but I personally doubt it. I do not believe that our island is big enough in its interglacial state. The migrations of 2 million wildebeest, joined by zebra and gazelle, from the Serengeti plains to the hills of the Masai Mara and back is a 1800 mile round trip. It is also, like the many other long distance ungulate migrations in Africa, driven by the rains. The vast animal movements are to track and take advantage of rainfall as it brings the low nutrient soils to life. In general we don't tend to see these scales of movement in temperate areas. In dry places, yes. In cold places, yes. But in mild and wet places, not particularly. Those who lived here probably stayed here year round. Our grazers may have moved from uplands to lowlands as winters came and went or from wetlands to drylands as flood waters rose or receded but I doubt their movements were as elaborate as those witnessed across Africa. From this perspective HPG in Britain is not quite the rekindling of an old relationship between stock and vegetation that it is in other areas.

There is, however, another type of interaction that possibly did play out between our grazers and grasslands in Britain and one that potentially can be mimicked by HPG. Chris Geremia and colleagues have been finding in Yellowstone that herds of grazers seem to have the power to create their own seasons; to trigger, through their grazing, a green flush of grass that is more often associated with spring growth. The bison that graze Yellowstone seem to have sufficient power to create their own 'green wave' as it has been termed, grazing heavily enough to encourage grasslands to flush without having to undertake large migrations.

The difference, once again, comes down to scale. These areas are small within the matrix of all of the other habitat types unlike the large scales on which migration mimicking HPG tends to be applied.

There is no doubt that encouraging grass to keep on flushing, by using HPG to mimic bison impact, supports far greater levels of stock than can be reared from the same ground otherwise. And by weaving this type of land use through less intensively managed areas we can add diversity into landscapes, creating areas of 'disturbed novelty' within a larger framework of stability.

Holistic Management is not a cure all but it points the way to reweaving the relationship between grazers and grasslands and restoring the diverse habitats that they can co-create.

What About Methane?

I have had many conversations about methane as its fame has risen in recent years and they generally run along similar lines. Methane is often singled out as a potent greenhouse gas and examined on its own, devoid of the system that creates and destroys it. I want to take a moment here to put methane back into its context and examine the system as a whole.

Methane remains a central part of the discussion around climate change being the third most important greenhouse gas, after carbon dioxide and water vapour. It is very potent, having an impact of about 34 times that of carbon dioxide over a 100 year period, however, it degrades in the atmosphere in about a decade, unlike carbon dioxide that can last for 1000 years.

Methane can be split into two groups, the biogenic methane produced by living organisms that tends to make the headlines, and the fossil methane which is often kept under wraps.

Biogenic methane is released by anaerobic respiration meaning that wherever matter is broken down without oxygen, methane will be emitted. This includes inside the digestive tracts of ruminants and across our wetlands, peat lands and sea bed. Even trees give off methane under certain circumstances as was recently discovered by researchers from the Open University studying waterlogged forests in the Amazon. The gasses that pour from holes drilled into such trees can literally be set on fire like a Bunsen in a school lab.

Studies on the effects of the quantities of greenhouse gases produced vs. those sequestered by trees, wetlands, peatlands and grazers are many and complex. In general trees and waterlogged soils have a positive climate impact, although much more research needs to be done to understand the balances more deeply. Grazers also seem to have a positive impact overall on greenhouse gases with the methane that they belch out cancelling out around 25% of the impact of the carbon dioxide removed by the pasture that they graze. Figures vary greatly, however, and most studies have focused on conventional systems where cattle graze ryegrass pastures. Simplified ryegrass pastures are poor capturers of carbon with diverse swards capturing 500% more carbon than monocultures. What is more, diverse swards containing tannin rich herbs seem to lower the quantity of methane belched out by ruminants further.

If we think of the vast herds of native herbivores that once patrolled earth, and all of the methane produced as part of their digestion, appropriate quantities of cattle, deer and ponies grazing similar landscapes and engaging in similar processes today cannot be the prime culprit. As is so often the case it is not necessarily as clear cut as cattle, sheep and deer produce methane and so are bad, it is entirely dependent upon the system of which they are a part.

Fossil methane is methane that is trapped within the earth and it can leak out of tears in the earth's crust; where natural gas and oil seep out, where gas bubbles from hot springs or where volcanoes rumble, for example. Our search for fossil fuels has led us to disturb these underground vaults releasing vast quantities of methane into the atmosphere by accident.

Fossil methane is a very different beast from its biological counterpart just as the carbon released from burning fossil fuels is a different beast from that exhaled by animals (not chemically different I should point out). The key to tackling both carbon dioxide and methane emissions has to begin with not allowing fossil forms to escape in the first place and then focusing on restoring earths cycling abilities to get the rest of them back where they belong before we start banning animals from breathing or belching.

Disturbance and Diversity

After 5 years I had left the farm and was renting 10 acres of land high above a wooded valley, right on the edge of the Peak District. The narrow road, moss bubbling down its middle, wound its way up from The Clatterway far below through a wood on the move slithering and sliding on scree slopes, falling ever downwards in slow motion. The road climbs higher and breaks out near the top of the tor to reveal a rusting tin roof beyond which my sheep graze. Theirs is a land of ravens, hairbells, twisted ash trees and wind.

Over the long hot summer of 2018 the vulnerability of our food system to drought was brutally exposed. Ryegrass pastures failed to grow, their short and lazy roots sat and waited hopefully for rain that did not come, encased in bone dry soil. This lead to a wide spread shortage of fodder; Swedish farmers bought all of the forage that a local famer could produce and cattle waited in empty fields to be sent to the abattoir, unable to be fed. The simplified farming system had jammed after only a couple of unexpected weeks. I watched as around me the emerald fields were put on pause and I apprehensively monitored my own grass.

We were already overstocked and I had to abandon any rotation as all but one trough dried up, the ever dependable spring had turned out to not be so dependable as farmers further up the hill drew off ever more water. My grass, however, trundled on alongside its motley crew of trees, herbs and scrub. By the end of the year I had some of the greenest fields going, and fattest stock, despite my thin alkaline soils as the rough and tangled mess of pasture showed its worth. Swallows still dived over head, a bonanza of bees and butterflies bustled through my thistles and my sheep contentedly chewed. Diversity won hands down that year.

If 2018 tested the resiliency of the Dales to drought, 2019 tested their resiliency to flood. From midsummer on it rained and rained. Almost every day water drifted from the sky, not in a torrential downpour but in a steady weep that seemed never to end. The river swelled, the roads ran like streams and I grew accustomed to existing within a shell of green flexothane waterproofing. Water eased up through the grass to sit in sullen pools over the valleys and the hills shook off their water, carving deep gullies as they went. Once more the mirage created by the bright glow of ryegrass faltered. The thick and sticky soils, uninvested in by their ryegrass roots and poorly covered by thin shoots, started to trickle away to collect in the slumps in fields. No one could get on the land to fertilise fields or plough under the stubble of the summer's harvest, let alone plant next year's corn. Those who did succeed in ploughing exposed the belly of their soil to the softest but most persistent of onslaughts that gently smoothed away their top soil down into ditches and onto roads.

The land had become laden, the river cradled her maximum volume of water, and then the heavens opened. Within 20 minutes the water had inched over the Derwents banks, trickled through fences, gushed down roads and plunged into houses, shops and barns alike. I lived on the opposite side of the valley to my sheep and I picked a careful path to see them winding through 'road closed' signs and puddles almost too deep for the truck as I snaked my way up the back roads onto the opposite tor through the haze of rain. From the ridge I stopped to gaze out over the suddenly alien landscape, a new sea shimmering between the hills under a still leaden sky. A farmer stopped beside me. He leaned out of his truck window, smeared in thick mud, an old hat pulled low to his wayward eyebrows and pointed to the top of some farm buildings now engulfed in water. 'See that', he said over the grumble of his pickup, gesturing to a small black shape next to a corrugated barn roof 'That's the top of a stack of five hundred bales that is'.

My sheep were fine, miserable and wet, but mud free and still chewing. My diverse pasture had handled it again, although, walking across the dip between two shoulders of hill, I noticed that we seemed to have acquired a new spring. The turf was loose and swaying above a crystal clear ball of water tumbling round and round in the mouth to a newly created soil underworld below.

The next day the flood waters receded and I went for a walk along the river. One of the 500 bales was perched in a tree metres above the normal level of the river and about a mile downstream. I wondered what next year would bring and how long it would take for people to start thinking about 'climate proofing' our food systems.

In an Eternal Land

In Britain when we think of productive trees orchards are what typically spring to mind; old gnarled apples that reach over springy turf, perhaps festooned in mistletoe and maybe home to a flycatcher or two looping from a branch over summer. The other familiar haunt of our productive trees is the hedgerow where crab apples and wild pears cluster between lines of sprawling sloes, hazels and hawthorns yielding a bounty of tart fruit each year.

It is not surprising that our fruit trees do well in sun drenched orchards and hedgerows because they largely evolved in scrambling scrublands where abundant light allowed their fruits to ripen.

Regenerative farmers have kept the idea of our fruit and nut trees as those that emerge from scrub or exist in sunlight copses but they have redesigned where this scrub occurs. They have taken the humble hedge and organised it, transforming it into a regenerative food production system of tree strips interspersed with alleys of pasture or crops. Alleys and tree strips can be laid out in more or less any arrangement to best bind landscapes together; they can be laid along the contour to catch water and build soils, they can stretch from north to south to best harness the sun's energy or they can be spaced to funnel people, animals or pollinators across the landscape.The trees planted could be fruit or nut trees for human consumption or forage trees such as willow, ash or lime for stock to eat. They could be species planted to shade stock or crops from the heat of the summer sun or to shelter them from the blast of winter gales. They could be trees grown for fuel or willows grown for weaving. Any combination of trees grown between or around other farming activities is given the term agroforesty, a term that in my opinion highlights our divided view of the world by its very existence.

Agroforestry systems stabilise landscapes by holding within them a massive diversity of species, both domesticated and wild. They not only stabilise systems across a landscape, however, they also do so through time. Vegetables, fruit bushes and trees have very different life expectancies; most vegetables are in the ground for less than 6 months, a blackcurrant might live for maybe 20 years, an apple perhaps 100, a mulberry 300 and a sweet chestnut up to 1000. This means that agroforestry landscapes are built to last and offers farmers a way of reweaving diversity and stability into landscapes in the long run.

The different life expectancies of our productive trees and shrubs, combined with yearly waves of planting, can build up a rich tapestry of productive scrublands and woodlands at different stages of development. A newly tilled bed can be sown to a crop of oats and while they grow soft fruit bushes can be established down the centre of the plot. In the years that follow fruit trees can be added to grow up through the soft fruit eventually maturing to tower above and shade out the cropped ground. Grazed pastures can then sweep in healing the soil and fixing nitrogen for the maturing currents. After years of supplying succulent berries the soft fruit will come to the end of its productive life leaving behind it a line of tall trees in its place. These trees can continue their relationship with the pasture or they can be joined by younger trees and bushes planted about their feet.

Forest gardens have become something of a craze over recent years within the permaculture world. A forest garden is essentially a collection of productive plants that mimic a young woodland with a ground layer of edible herbs, clumps of a wide array of useful shrubs and a patchy canopy of nuts and fruits. These aspiring woods can be planted around tree strips to flesh them out into wider pools of trees. Over time they will mature into older woods and their canopies will start to close. Mushrooms could be grown in the cool and damp gloom between mature trunks or straight timber trees drawn up through them. Grazers too might return one day to transform the woodland floor into herb rich wood pasture leaving only the long lived canopy trees towering above them, perhaps recreating our ancient orchards home to mistletoe and flycatchers. The tallest and longest lived trees of all, the nuts, would one day stand bent and twisted above the decedents of those livestock and perhaps still home to a territorial mistle thrush. These veteran nuts would have found their feet in the soft fruit alley, joined in the bonanza of the forest garden and outlived all others in the wood pasture. They might drop their fruits each year amongst cattle, sheep or poultry and fatten the last of the summer's pigs for the freezer. From their lofty positions the nut trees could look out over a landscape of alleys; to their offspring in their soft fruit scrub nursery, to their fruit bearing neighbours, to their older progeny peeking out from above the forest garden canopy, to their own cohort still forming the ghost of an alley snaking away across the hillside. Connected to all by a secret underground network of filamentous fungi. These trees could have stood for hundreds of years, sucking in and pumping down carbon at its peak atmospheric content, sheltering those that followed in their wake, remembering the pests and pathogens of the past to forewarn their offspring and steadily nourishing the landscape as a whole.

When we see the phases of a habitat as one long dance we see how all regenerative agriculture blends together to restore landscapes of continuity across time and space and how human ecosystem engineering can co-create these landscapes to foster future human life.

These are the landscapes that the ecosystem engineering, keystone species *Homo sapiens* can co-create.

Return to the Sea

English geographer Jay Appleton published a book entitled 'The Experience of Landscapes' in 1975. In it he introduced his 'prospect-refuge theory'. Appleton suggested that people prefer to see without being seen. We like the panoramic views of a hilltop, a roof terrace or an open field but we also like the security of a cleft in the rock, a comfortable sofa or a thicket of trees. Appleton suggested that this deep attraction to landscapes that are both open and secure stemmed from our evolutionary past on the African savannah, a landscape that offered us opportunities for prospecting as well as secure refuges that would have aided our survival. It is likely that we evolved alongside such environments and researchers today have confirmed that, all around the world and regardless of culture, people are most attracted to, and relaxed in, such places.

It is also interesting that humans have created these landscapes around the world and through time. Examples include the edible grasslands interspersed with oak trees left by ancient inhabitants of the Fertile Crescent, the expansive grasslands and productive trees maintained by the Native Americans and the English parklands of deer grazed pasture interspersed with veteran oaks. People seem to create these landscapes automatically, through long dances of co-evolution, by following instincts of what is pleasing and by designing landscapes that will feed us sustainably. Modern agroforestry systems are their latest incarnation.

Many people around the world are developing the first alleys populated by the first fruit and nut trees, grazed by the domesticated decedents of native animals; yielding a smorgasbord of fruits, nuts, cereals, vegetables and meat.

I had taken up the position of Farm Manager on one such farm during its first season of operation and I arrived at the local train station on a cool, clear April day. I was met by one of the owners and we set off towards the farm; weaving through mature, deep green plantations as the roads became wilder and the tarmac gave way to gravel. Round a final bend and there it was, a red roofed, pale walled farmhouse and behind it a cluster of barns falling away to fields fringed by the heavy shadow of deep green conifers beyond. The owners hadn't been there long and the land and buildings had the lazy, settled feel of a landscape that has been biding its time. That was not to last. Over a couple of weeks the team assembled and we started on the grand plans for the place. We got the veg garden up and running, laying metre after metre of wet cardboard on top of which we piled tonnes of compost and woodchip hauled with a wheel barrow from a small mountain tipped near the road. As the instant vegetable garden emerged we began on the next big project; the keyline tree strips. One morning an agricultural contractor arrived and grinning, although slightly bemusedly, hooked the Keyline plough up to his tractor. The intern team had marked out a series of parallel lines across the landscape that would one day allow water to settle into the ground evenly, the ridges and valleys alike, just as Yeomans had suggested. The tractor revved up and off the contractor set, up, down and round, combing the landscape with tiny ruts like a bizarrely decorated cake.

Once the water relocation system had been established the tree planting could begin. Hundreds of apples, pears, plums and cherries surrounded by a sea of black currants, gooseberries, hazels, raspberries, honeyberries and Siberian pea trees started to snake their way across the fields. Over the next few days we endlessly planted through the sleet and drizzle until the finished pattern of the landscape finally emerged and we could get rolling with livestock. As the member of the team with the most livestock experience it soon became my role. Setting out fences we began rotating the cattle between the tree strips and moving the chickens after them. The hens, safely stowed in their mammoth 'eggmobile', trundled over the fields turned out every day onto fresh pickings left for them by the cows. The broilers, chickens raised for meat, also engaged in this steady dance being dragged one section at a time in their movable pens to peck at, and shit on, grasses new. Together the poultry deposited their grain based diet as fertiliser over the fields which glowed bright green in their wake.

The farmer was following the example set by Joel Salatin, the famous and self-proclaimed 'lunatic farmer' behind the world leading Polyface Farms in Virginia, USA. Salatin moves his cattle daily and his layers follow three days behind to eat the maggots that have by that time been deposited, and grown to admirable size, within the pats. The chickens get some free protein and the farmer reduces the number of dung flies that hatch from the fields.

Moving the miniature raptors on each day, and watching them tear through everything that they came across, reignited the old debate in my brain. The farmer's desire to profit from the free foods of nature slogged it out with the conservationists desire to protect for the future.

I had never encountered such diverse fields before, other than in a few discreet nature reserves, because they quite simply no longer exist across most of England. These fields, however, harboured literally hundreds of species of plant; a bonanza of grasses as well as bellflowers, hairbells, oxeye daisies, field scabious, self heal, sorrels, stitchworts, strawberries, vetches and vetchlings. They also had a thriving population of yellow rattle. Yellow rattle is a plant that I have a long history with. It is one of a conservationist's best friends and many farmers worst enemies because yellow rattle parasitizes grass, it leaches the grass's nutrients hobbling their growth and allowing herbs to flourish beside them.

Through, below and above this jungle of diversity coursed an army of invertebrates; grasshoppers larger than any I had ever seen, butterflies and moths of all descriptions, beetles, worms and spiders. There was a colossal quantity of life that had adapted to the late cutting of hay year on year and I wondered what the sudden arrival of an army of chickens would have on this abundance as the keen eyed hens picked out and gobbled down this wealth.

I started to gather information on the numbers of dung beetles and maggots in dung pats of different ages, the identities and quantities of life searing through the sward prior to and after the passage of the poultry and the roles that these species might be playing in the grassland food web. How many invertebrates could be consumed over what time period before their rate of breeding, rapid as it is, was outstripped by predation? Was there a stocking density or percentage of ground cover needed to ensure that sufficient adult invertebrates survived to breed? Was there an optimal level of 'grazing' of the invertebrates by poultry, as there is grass by cattle, to avoid local extinctions? How long did dung beetles require to gather their dung, lay their eggs and vacate before pats were ripped asunder by dinosaur talons to ensure the next generation of dung decomposers? In other words, what impact would the arrival of an ecosystem engineering, regenerative farmer allied to keystone livestock have on a landscape? My curious probing, as usual, raised more questions than it answered and those questions would tip me back head first into the world of ecology and ultimately, pave my way back to the centre ground between farming and conservation.

In Search of Balance

Farming, especially over the last couple of hundred years or so, has been characterised by control. People have willed highly productive monocultures into being and then we have used our considerable might, backed by fossil fuels, to try and keep them in a simple yet productive state. We have spent decades beating back 'diseases', identifying and combating 'weeds' and targeting and exterminating 'pests'; all to maintain an ideal of civilised, controlled super abundance. This is a task worthy of Sisyphus if ever there was one; forget pushing a rock up a hill, battling the natural processes of the world, now that's an endless slog. Regenerative agriculture abandons this battle and chooses to look instead in the opposite direction. Not at how to get rid of disease but at how to build health.

Each outbreak is an imbalance, a discrepancy between the population of one species in a food web and those of the other species around it, which causes the food web to flex as a whole. As the system moves through and processes the imbalance it gains the ability to steady itself against that imbalance in the future. For example, a reseeded pasture grazed by cattle may have chinks in its turf into which ragwort seeds settle. This ragwort may grow and its population swell year on year but, as any resource swells out of balance with the rest of the system, it is likely to attract the attention of onlookers searching for a meal. Cinnabar moths are ragwort specialists and a ragwort bonanza means a cinnabar moth bonanza shortly afterwards. Once the cinnabar moths have found an area of ragwort they will probably remain. There will most likely be ragwort each year in some nook or cranny but there will also be cinnabar moths, the populations of the two working together towards balance as long as no major disturbances occur.

Systems can be viewed to a certain extent as having processes analogous to our immune responses and, just like applying antibiotic too soon, removing a pest, weed or disease can rob the system of its opportunity to learn how to combat the imbalance in the long run. Systems 'learn' to recognize and respond to the species within and around them; whether they are bacteria invading the body from a cut in the skin or ragwort blooming in a tear in the grass. Once the system has learnt how to deal with the new arrival it must remember that information and, just as our bodies store information on the pathogens that we have had contact with ecosystems store the species within them. Pests need to be present within the system for the whole growing season to keep their pest predators fed and weeds need to remain present to feed the animals that eat them. It is a very counter intuitive idea but one that makes perfect sense; in order to avoid plagues of pests, diseases or weeds regenerative farmers need to make sure that they are *always* present. They need to ensure that their systems don't forget how to respond.

It is a careful line to walk, one of allowing systems enough free will to find stability and destabilising them sufficiently to encourage them to offer up a bounty of human food that they would not produce otherwise.

My own experience with this method of farming has been fairly successful; in supporting sufficient quantities of nettles to harbour caterpillars, in retaining docks to feed dock beetles and in keeping enough intestinal worms to prompt sheep to maintain immunity. It has not worked for thistles, however.

I have spent countless hours scything, slashing, strimming and topping thistles in an attempt to reduce their number, playing my role as keystone omnivore and harvesting abundance until I was covered in sweat, sap and thistle spines. For a few years, however, I decided to experiment with what would happen if I just let them go in certain areas. I would not recommend this as two of our thistle species are defined as 'injurious weeds' under the Weeds Act of 1959, however, I decided to cordon of an experimental patch as far away from my neighbours as possible and give it a go anyway.

In no time at all a silver hue of spiky leaves hung above the pasture like a light mist rising slowly upwards and by July the field was lost below a purple haze that hummed with bees and hoverflies. Peacock butterflies rose in plumes from the sea of flowers, betraying the movement of sheep below the surface. Retrieving the flock became spiky work and what was worse was that the sheep brought the thistles with them, lurking in their fleeces, making even simple jobs very prickly indeed. My experiment was not going particularly well, even from my perspective, and it was drawing a lot of unwanted attention from the neighbours.

By the back end of summer I was starting to think that all of the chopping done in previous years was defiantly worthwhile. My thistles had grown abundantly, had their flowers more than adequately fertilised and now, as I cast my eye across the field, their seeds were beginning to be released to the wind. The small square was a tapestry of green leaves, purple globes and downy white seeds. I started across the field to assess the crop, and track down a missing ewe, when a charm of goldfinches and greenfinches rose from the jungle ahead of me. The air was thick with frantic wing beats and streaks of red, gold, yellow and green for an instant before they descended once more into plumes of white seed. I watched a goldfinch adeptly swinging on a bowing thistle head, tugging at the feathery down. Expertly it twisted the fluff in its beak to extract the seed before tugging another, and then another building up a snowy beard below its chin. 'At last', I thought, something had arrived to capitalise on the abundant food supply.

As summer drifted into autumn the charm of finches became a regular feature of the thistle field and as autumn descended into winter they melted away once more, like the thistles themselves. Life recoiled from the creeping cold and my barns filled up with thistle fattened butterflies, clinging like black leaves to the undersides of the timbers.

Winter gave me time to think, leaning against the rotten door frame and gazing out over a frost hardened landscape, I imagined the networks of brittle thistle roots that lurked beneath. The finches had done an admirable job of clearing away the thistle seed but this would only regulate how rapidly and how far the thistles spread. I had creeping thistle in my bunch that, as the name suggests, infiltrates fields below their surface. Its seeds are not the problem, in fact they are often sterile, it's the roots you have to worry about and I could not think of anything that would stand in their way.

Leaning in the barn doorway, where the warm smell of animal bodies and hay mingled with the clear and crisp winter air, I wondered what might get rid of thistles and then it occurred to me; woods might. Grassland is not meant to stay as grassland for ever more, landscapes are meant to move. Thistles do a fantastic job of creating dense forests of spines below which stock are often unwilling to graze. Their protective embrace might very well offer enough protection for other, woodier, plants to sneak in and get established.

Over the brow of the hill I was conducting another informal experiment, this time with hawthorn. Sheep get a very bad press in the environmental world for being the destroyers of woodland and yet my observations did not align with this at all. I had grazed my hawthorn experiment hard, year round, and my sheep had nibbled the leaves off the small bushes numerous times but they just would not die. Whether secretly plumbed in to the underground mycorrhizal distribution network or just incredibly tough my grazers could not get rid of them from their grassland. It looked like the hawthorn would eventually take over. There was even a young ash sapling already springing from a more established thorn bush straddling a dry stone wall. Birds would inevitably perch in the young tree and equally inevitably excrete their cargo of berry seeds, stocking the thorny patch with possibilities. Small mammals too would scurry below the outstretched arms of thorn, perhaps stockpiling a stash of nuts for the winter in some hidden and out of the way place. Over time thorns, briars and trees would establish and spill from the thicket transforming the grassland into woodland and it looked as though my sheep would not be able to stop them. Perhaps thistles laid the ground work for this process to begin and perhaps it was woodlands that cleared the thistles from landscapes once more? I had no way of knowing; on rented land I had already pushed my luck enough.

Everything has balance and everything moves in cycles, some short enough to see and others so long that they pass us by. Regenerative agriculture engages in these cycles and aspires to work with them to steer agroecosystems towards balance, stability and diversity.

Mimicking What?

Across permaculture, agroecology, regenerative agriculture, holistic management and many other progressive agricultural techniques the term 'mimicking nature' springs up. It is one of the strands that they all share but what does it mean? To mimic something means to imitate it and the implication can only be that we, and our farms, are not 'nature'. The farm must be a copy, an imitation, a mimic. The wild is seen as teacher, as aspiration, as perfection; but also as discreet from farming. The term reveals that even the most alternative and progressive farming systems are built on a belief in the divide between people and the wild.

Alternative farming techniques have come amazingly far in threading our word back together; in binding water back to land, in merging soil and plants, in seeing grazers and grasslands as two sides of the same coin and in planning for the next 1000 years. But they generally do not see people and farms as a part, and a beneficial part, of the wild. They do not tend to see farmed systems as diversifying pools within a whole but as delineated and corrupted versions of nature. Alternative farmers allow nature in and learn from her teachings but they still hold a space for human occupation, view farms as areas driven by human will and use a language of difference. They still label some areas wild and others cultivated; they call some plants weeds or companions and some animals pests, game, pollinators or vermin. We see ourselves as 'land sharers' but even the word *share* implies a divide because there must be two parties in order to share something in the first place. Bound up within our language and farming are old divisions and we still harbour a disconnected world view.

Part III

A Whole Future

Chapter 6

Integration

It was hot. Too hot to do anything and I slumped down amongst the sheep high above the valley. They had also concluded that it was too hot; some had nestled themselves below the old crab apple, others had tucked themselves away next to the dry stone walls or embedded themselves in the fringes of scrub. I had chosen a spot in a slight hollow next to an ant hill adorned by a crown of five perfect rabbit dropping. I lay back and listened to the clatter of swallows high above me, to the buzzard drifting down the valley and to the faintest scuffling coming from the sward. Miniature flowers of all colours bobbed their heads as they protruded from the thin grass, the limestone bones of the earth breaking the surface, their rounded shapes like a pod of dolphins. A week before I had been walking through ancient Atlantic oak woodland on Scotland's west coast seeking signs of beaver. Their less than subtle munching of whole trees and the large and messy piles of sticks that they called home far easier to spot than I had dared hope. Their flooded woodland and network of paths obvious despite their relative newness to the place and the place to them. My mind wandered to Yellowstone, to the wolves and to their beavers. To the expansion and influx of life that the two had conjured into being. To the song birds and the amphibians that flooded in behind them.

Later that day the heat of the sun still poured down, on my back this time as I knelt between a line of neat cabbages and an untidy sprawl of flowers. Buzzing filled the air all about me and a thick slick of butterflies flopped haphazardly over head. As I peeled back the layers of chickweed interspersed with fat hen a devil's coach horse beetle scuttled away to hide, unwittingly, under my shoe. I gave it a moment to rectify its mistake and looked about me. There was a thick bank of nettles washing into the sweet peas, thistles loomed against the fence and a sea of willow herb and enthusiastic leeks leaned near the pond. This garden harboured more insects than I had seen for years, it was also weedier than it had been in years. The swallows too knew that this was insect capital, darting over head they were clearly focusing their attention on this one acre of messy vegetables and flowers out of a hundred adjacent acres of neat ryegrass, trimmed hedges, dry stone walls and dairy cows. Within this context my previous musing returned to me, wolves and beavers, song birds and amphibians. We had tried to encourage the latter pair for years, there was definitely plenty for them to eat here with slugs that nipped broad beans off at ground level each spring and caterpillars that turned leaves to lace over summer. Looking about me I expected to see the caterpillar modified cabbages but, despite the abundance of white butterflies, I could not spot a single one. I wondered if we were finally reaching some sort of balance and the bird population had wised up to the buffet on offer. If we had wolves or beavers would we have more birds and amphibians? Would our veg grow better?

Sitting in that sun drenched garden, mindlessly scooping up weeds, an idea floated into my head. Had humans, by removing all of the top predators that had presumably been regulating our ecosystems from the top down for centuries, actually sculpted a system that was only controlled from the bottom up? Was the abundance of food now the thing that largely determined animal population levels and if so, the endless attempt of people to grow the maximum amount of the most succulent food was really just a way to grow a lot of things that ate it. Far from threatening farming, could rewilding and the addition of things long lost, be its salvation?

Looking back on it now it seems a ridiculously obvious brain wave, of course it can; but sitting there amongst the jostle of life it felt nothing short of revolutionary.

An Island in the Sea

There is so much resistance to seeing the world as one endlessly changing whole that few people seem to have pondered how rewilding and farming might fit together and support one another in a single landscape.

The three strands of rewilding are cores, corridors and carnivores and farming can support rewilding with all three of them. The first two strands, the need for cores and corridors, were born from the theory of island biogeography, as we discussed earlier. The use of this theory to guide reserve layout was a stroke of genius which allowed massive strides to be made in determining why species did not persist for ever more in protected areas. As useful as this theory has been we must not lose sight of the limits of its applicability. Whilst islands do exist in an environment which is so different as to be uninhabitable by land based life, reserves do not. The land around and between reserves is most often farmland which has different 'grades of permeability' based on how it is managed; in other words it varies from very sea like to very terrestrial.

As we have seen our wildlife is well adapted to scrub, copses, wetlands and grazed clearings and so there is nothing massively alien about an organic landscape of shelterbelts, hay meadows and grazed pastures or a regenerative landscape of tree strips, grazed cells and irrigation pools. Farmland can therefore act to massively expand the area functioning as the core or corridor of a reserve allowing our wildlife to amass greater population levels and migrate more freely. In its present form, however, farming is more akin to the sea and tends to restrict wildlife movement and squash populations back into their reserve sized boxes.

Although this makes a lot of intuitive sense there has been little work done on the permeability of different methods of farming in Britain, and non that I am aware of on the permeability of regenerative systems compared to industrial ones. Some research has, however, been conducted in Southern Mexico on bird movements through coffee plantations. Coffee can be grown in one of two different ways, either as shade coffee below a canopy of forest trees which is the more traditional approach, or as sun coffee without the protection of a canopy which results in higher yields of poorer quality beans. Research reported in the brilliant book 'Natures Matrix', carried out by the authors, monitored coffee plantation use by a forest bird, the highland guan. The highland guan will happily pass through the more ecologically intact shade coffee plantations but will not pass through sun coffee plantations. The type of farming practised determines whether or not the land can act as a corridor for birds moving between sub-populations in habitat fragments.

Currently in Britain our nature reserve islands are managed by selective felling of areas of woodland to create glades, the steady coppicing of understory trees to form a shrub layer, the loose grazing of grasslands over winter or the late summer cutting of hay. They are islands of traditional, hopefully organic, management in a sea of modernity. If our agricultural landscapes were created by regenerative farmers and woodsmen, however, in the absence of wide scale tilling and agrochemical use, there would cease to be a large difference between the reserves and the surrounding farmland sea at all. The land area available to be used by our wildlife would increase massively and the corridors that link up our reserves would no longer be required. The whole landscape would be one interconnected, wildlife rich, whole.

This, however, does not meet the desires of many who yearn for wilder, self willed areas on our island and provides no place for 'wild incarnate' species to dwell. For this we must go further.

We currently use around 20% of the British landscape to produce enough wheat and barley to feed everyone about 3000 kcal per day. Another 7% of our island is used for our cities, towns, roads, houses, gardens and parks. It is, of course, impossible to know for sure but it is highly likely that a regenerative system of tree strips, alleys, cereal plots and veg beds could make a significant contribution to our food needs in the future, leaving potentially vast areas for wilder uses. The landscape could turn from a simplified tapestry of chemical doused land with fragments of traditional management bound within it to a tapestry of regenerative human engineering surrounded and permeated by an ever shifting tapestry of rewilded life. Within this context our engineering would once again create the disturbance within the stability helping to raise landscape diversity once more.

Regenerative agriculture minimises the degree of difference between the 'island' and the 'sea' but it also offers a solution to the final part of the rewilding jigsaw; carnivores.

We have very few large wild grazers left in Britain and we have no large predators which is why places like Knepp rely on cattle, ponies and pigs to cause disturbance and humans to regulate their population levels. People remove a percentage of the grazers each year to replicate an aspect of predatory impact and, if this is not done, animals will run out of food and starve to death as the managers of the Oostvaardersplassen discovered. This ties rewilding, for now at least, to managing herbivore populations and playing the role of the wolf ourselves. In other words it ties rewilding to farming. Although managers can go onto a site and remove a number of animals every year, they cannot create the dynamic 'landscapes of fear' that wolves create. Those relationships can only form when there are repeated interactions between predator and prey that build up over time. They can only occur when people and grazers live in the same landscape and dance around one another for generations. In Britain we have a long history of this relationship in two guises, woodsmen and shepherds, and both are allied with the descendent of the original creator of landscapes of fear; dogs. Woodsmen were the quiet threat that lurked in woodlands making everything from charcoal to hazel hurdles and turning the limbs of trees into all manner of items from chairs to bowls. These people carefully tended woods to encourage them to grow the materials that they required for their craft which were mostly drawn from coppiced trees. They felled small coupes in rotation and then guarded the delicate regrowth alongside their dogs. Between them the woodsmen, the dogs and the browsers built up tapestries across woodlands strewn with pools of wildflowers, butterflies, regenerating woodlands, grazed glades and fear.

Shepherds carried out the same role over open ground. They kept up a steady pressure on the herbivores corralling the livestock over grasslands, again often accompanied by dogs. The shepherds encouraged the stock to stay bunched and moved them as one creating a mosaic of disturbance across the hills and valleys as they went. Generations of shepherding wove a rich and diverse fabric from the threads of sheep, plants, people and dogs facilitating abundant grasslands, woodlands and wetlands to spill through landscapes.

Could people and their dogs not do the same again? Could we not fully take up the role of wolf and create landscapes of tree protecting fear if we wove our lives between those of the wild and allowed a deep connection to form once more?

We are charting new territory and there has been very little research, none in fact that I am aware of, into how human engineered landscapes can support rewilded ones but we need to start somewhere. To me it seems obvious that regenerative farmers can support rewilding by converting what is now hostile sea into something remarkably terrestrial and by shaking up grazer movements across landscapes. We need an active conversation around this topic, to lever open Overton's window and explore how agriculture could support rewilding in theory and practise.

Altering farming to enhance the value of land to wildlife is not particularly controversial. Suggesting that the relationship might work the other way around however, and especially that *rewilding* might benefit agriculture is slightly more contentious, especially amongst farmers who often feel under threat from wilderness enthusiasts as they eye up their farms and fit them for size for wolf packs. All good partnerships are equal, however, and below we will explore if, and how, rewilding can benefit agriculture. Would more beavers result in healthier vegetables?

A Cradled Turmoil

It's probably a bit early in the game to guess what impact beavers might have on vegetables but the general principle, that a diverse and stable matrix of wild land boosts farmland productivity, holds water.

As we have discussed, an agroecosystem can only function well when complex populations of weeds, pests, diseases and their natural predators build up and regulate one another over time. Farmland tends to experience more disturbances than 'wild' land, however, as people mow fields for hay, harvest crops or till areas; radically altering them quite quickly. These disturbances often cause species to go locally extinct leaving pools of unused resources in their wake. A recently tilled strip of land, for example, has an abundance of untapped resources and the first species to reach, and make use of, those resources will do very well. Often those species are the ones that can travel great distances to find new resource pools and reproduce very rapidly once they get there; we tend to label them weeds and pests because of these traits. A new bed of broad beans will likely be found by blackfly dispersing from their winter haunts that, being born pregnant and capable of giving birth to live young, will quickly multiply to take advantage of the resource pool. Ladybirds, one of the blackfly's main predators, have a much more laid back approach to life. They hatch from an egg, grow for a couple of months on an aphid diet, metamorphose into a shiny spotted beetle and often do not even think about finding a mate until next year. Their population growth rate can nowhere near match that of the blackflies and the pests run riot while the ladybirds steadily plod away.

Our broad bean bed does not exist within a vacuum, however; it is embedded within a larger landscape that can house millions of blackfly predators, if only they can safely and speedily reach the blackfly bonanza. Intricate networks of small farms in close proximity to wilder land can allow ecosystems to re-form balanced food webs very quickly after a disturbance as species, such as ladybirds, flood in from nearby wild refuges to capitalise on the blackfly abundance. This can only happen, however, if our pest predators have stable areas to radiate from and safe lands to move through; it can only happen if farms are threaded through the wild.

While I was working as farm manager at the regenerative agriculture farm, I had wondered how many cow pats needed to be protected from the chicken patrol to ensure that dung beetles, and other dung fauna, did not go locally extinct within the field. In the current context this becomes the farmer's problem, most farms are small and isolated and the one on which I worked was set in a sea of conifer plantation that did not harbour high levels of dung beetles or a diverse array of pest predators. The ecosystem engineering, regenerative farmer interacts with the farm but beyond the boundaries of the land the wilder system does not exist in a healthy and dynamic state, it is not capable of housing and offering up an array of species from which disturbed food webs can be rebuilt.

A human's role is to disrupt, shake up, diversify areas of land within larger, stable landscapes, but when this is done without the surrounding matrix of the wild it opens the farm system up to localised extinctions. Regenerative farms have to host the disturbance and *also* maintain the stability required to support pest predators, pollinators and all of the rest of a diverse food web.

But they shouldn't have to. Engineering farmers shouldn't need to engineer the area of difference *and* retain the stability around it; the wider, wilder landscape should be healthy enough to provide those species just like rain, oxygen or sunlight. These should be things that spring from the wider landscape, free of charge and for everyone's benefit. And these things do spring from a rewilded matrix. Productive systems need to be imbedded within stable, wild ones for maximum regenerative productivity.

Again, scientists have yet to start to uncover just what effect a rewilded matrix has on organic farmers embedded within it and so it is difficult to gauge what effect this has on farm productivity. There are, however, a lot of well researched examples of species spilling out of reserves into adjacent farmland and the positive effects that this has.

Research has been carried out on how far pest predators stray into chemically managed crops from wild margins and the idea of beetle banks has arisen from this work. Beetle banks are strips of tussocky grassland that run through arable fields and provide an overwintering habitat from which predatory ground beetles and spiders will spill to gobble up pest species. Pest predators such as parasitoid wasps will also use beetle banks. They scout out caterpillar hosts for their eggs and will use 'wild' caterpillars growing on native vegetation in the banks as well as 'pest' caterpillars on vegetables as they become available. The same can be said for ladybird or lacewing larvae feeding off both wild and pest aphids or any number of other pest predators. The quantities of pests that are captured by invertebrate predators is truly staggering with 7 spot ladybirds eating between 20 and 500 aphids every day depending on life stage and Britain's population of social wasps gobbling down as estimated 14 million kilos of invertebrates each year.

Pollinators too flow out of wild lands to visit the flowers of many commercially grown fruits and again farmers have already captured this effect with wildflower strips sown down the middle of fields of crops that require pollination. Even where honey bees are kept farmers often encourage wild bees because they are far more effective pollinators. Many wild bees have very densely haired bodies that carry pollen far more effectively than a honey bee's body and the diversity of different bee species also have different tongue lengths allowing them to make better use of specific flower shapes. Certain bees, like bumble bees, are also capable of something called buzz pollination. Pollen is an expensive thing for a plant to produce and so some plants, such as tomatoes and blueberries, have evolved to hold onto it tightly until a trusted delivery service comes along. Bumblebees are capable of contracting their wing muscles and directing the vibrations produced at the pollen securing parts of flowers, stimulating them to release their precious cargo. This can be heard as the high pitched, short buzz between the lower pitched hum of their flight between flowers.

Although the overflowing abundance of wildlife might be most useful to arable farmers or growers, livestock farmers can benefit as well.

One of the more challenging aspects of rearing livestock, and sheep especially, is the plethora of internal parasites that wish to make the stocks' gut their home. The adult parasite, living within the grazer's intestine, lays eggs that are deposited onto pasture in the animal's dung. The eggs must then run the survival gauntlet while they mature to the stage at which they wish to be reingested by the stock to complete their lifecycle. Dung is a whole community and food web in itself and life for a parasite in a well used pat can be tough. There are various fly species depositing eggs in the dung that hatch out into maggots, most of which feed on the dung and microbe slurry but some of whom prey on other maggots. Feeding off the maggots are a range of predatory beetles and their larvae that roam through the pat searching for prey. There are also the dung beetles themselves, over 60 species in fact, that tunnel through, or construct elaborate burrows below, the pat. Some even dig down deep into the soil and provision secret underground lairs with balls of dung for their offspring to munch through as they develop. All of this tumultuous life, with its eating, burying, turning and distributing lowers parasite egg viability substantially. The teeming pats attract birds that peck through the dung scattering the dung still further and flinging parasite eggs everywhere. Exposed, and bereft of the moist and cool cow pat interior, the parasite eggs and larvae dry up and die.

The good thing about all of this dung decomposing food web; the flies, the beetles and birds, is that the vast majority of them are strong fliers and are more than happy to venture into farmers fields from rewilded lands beyond to lower stock parasite levels.

When I first started to think about how rewilding and regenerative agriculture sit next to one another and blend together I thought of the relationship as fairly simple. Permeable farmland expands the area that can support our wildlife and allows a free flow of species across the landscape and farmers benefit from that flow of species through their farms as stable food webs reform rapidly after disturbance. The more I thought about it, however, the more I suspected that the relationship was deeper than this. I suspect that, not only are wild lands providing the stability for farmed lands but farmed lands are providing the novelty for wild lands. In areas of human ecosystem engineering the different pools of habitat created support species that are not found in the 'background' ecosystems of the area. Farmers open the door to a wide array of annual plants and associated microbes as well as invertebrates, birds and small mammals that require bare soil or a high density of flowering and seeding plants to thrive. These species live within the ever changing systems that we create and then reproduce, stocking the soil with their seeds, cysts or eggs for next year and the years that follow. The dynamic nature of human managed areas supports generation after generation of plants, insects and microbes, holding within them a colossal quantity of DNA, recombining into unique individuals and building into new species and novel food webs. Whilst wild land provides the stability, I think that it's likely that human engineered land generates novelty.

The mixing pot of our dynamic engineering could very well be generating unique individuals and species and trailing novel food webs that respond well to soil disturbance, sudden defoliation or other large impacts. These species might then migrate back out into wild lands beyond to lie dormant in the soil until an opportunity for their growth presents itself. Could it be that regenerative farms, woven between rewilded lands, disseminate the evolutionary learning from disturbed lands to stable lands? Could regenerative agriculture add to the options of wild systems and enhance their resilience in the face of natural disturbances, perhaps even extreme ones caused by climate change, for example?

Ecologically speaking, wild stability can never be separated from disturbed human ingenuity; they are one and the same and function as complete, whole landscapes from which the amazing properties of stability, resilience and evolution spring.

Nouns and Verbs, Structures and Processes

At several points throughout this book, we have run into a common problem; that of language. We ran into difficulty with the distinction between woodlands and grasslands, wild and wilderness and again with native and exotic. I have so far skirted around the issue, however, in writing the first half of this chapter, language has become a constraining factor once again; this time in the form of wild and farmed lands. It is now time to grasp the bull by the horns and address the issue head on.

Language, in many ways, is the primary building block of the water in which we swim. It evolved as one strand of a toolkit that lead to who we are today. To understand just how foundational language is to how we interact with the world we have to step back in time once more, to a time when our distant ancestors first started to huddle about a fire.

When we harnessed fire we had, possibly for the first time, a focal point for the group; a central hearth that provided food, warmth and security from predators. According to Daniel Everett it is quite possible that fire provided the nucleus around which the group coalesced and community was born. With the new cohesion of the group and the additional calories offered up by cooked foods, our brains started to alter and language was most likely born.

This seemingly obvious set of changes was truly monumental. In order to develop both language and community a person had to be capable of imagining that other individuals have experiences. In other words a person needs to understand that others have perspectives and knowledge unique to them. This is called a 'theory of mind' and children generally acquire it at around four years of age. This theory of mind allows an individual to guess what someone else might be feeling or thinking and opens the door to language as a means of relaying one person's experience to another.

In order to relay verbal information people first required words. Most languages of the world have the same base words, things like wind, water, tree and woman, that were probably named fairly early on (interestingly most cultures also tend to have the same base gods; sun god, sky god, earth god). Over time these base words become the foundation of other words. Once the concept of 'woman' was established, for example, the idea of wife, mother and prostitute might emerge. From here concepts such as fidelity, adultery or betrothal may develop and from there words like cheating, divorce or spinster. These are words that describe external 'things' but they come attached to a variety of subtle emotions, connotations, assumptions and stereotypes that guide how we feel and what we anticipate. Once the base words and associations are created further words can be built upon them leading us down a rabbit hole of intricate cultural understandings. In this way, although most languages share a common underpinning of the words for the material world that surrounds all of us, very rapidly highly specialised ideas unique to the culture that created them appear forging many thousands of ways of understanding and interacting with the same world.

This 'Great Naming' is ongoing to this day and fundamentally divides our world. By naming a river we segregate it from its floodplain, the rains, the clouds and the sea all of which are a part of it but attributed to a different section of it, to a different word. By naming something we encapsulate it within the container of its word and assign attributes to it. The thing ceases to be what it is, with its own force, and starts to be what we ascribe to it, what we project onto it, what we assume to know about it based on its word container. The process of naming is the first step in gaining control over our world and in structuring the relationships of power that underpin it. In other words, naming something is a projection of our will and value system onto it.

These ideas have been developed and expanded upon within the LGBTQ+ world as people have pushed back against binaries of gender and sexual orientation ascribed to them. Grey areas, such as the ground that lies between or beyond the binaries of gay or straight, male or female, have been identified by people who have forged new words to give themselves existence. Gender queer, androgynous, gender fluid, nonbinary; all words to describe the liminal zone between the commonly recognized opposites of male and female.

As our language shifts, so too does the world that we live in. We can only talk in the words that we have, meaning that we can only describe the things in the world that are named and we can only imagine a world constructed of those same named entities. People used to live in a world of male and female and could only imagine a future based around heteronomative life. With the birth of new words to describe other ways of being, and other futures, people started to craft words to describe something else. With the creation of the word, our perceived world changed shape, followed by our real world with the appearance of gender neutral toilets, clothes that are described in terms of size and shape instead of gender and legal systems that acknowledge more than the gender assigned at birth of their citizens. The words that we use and the weight that they carry forms, in a very real way, the world that we live in.

Our language facilitates the continuance of our culture by recreating and reinforcing the world as the culture believes it to be and curtailing the description or imagination of anything beyond the confines of that culture. Our language fundamentally frames and guides how we see the world and English especially divides landscapes up and freezes them in time. An English speaker will have no difficulty in distinguishing between a river, a pond and the sea but there is no easy word to express the eternal flowing water that unites the whole world and so people tend not to see it. We have words like keystone grazer and hyperkeystone human but no word to describe the complexity of this relationship; that even though it is the hyperkeystone human that brings cattle to a region it is the ability of the cattle to sustain people in that landscape in the first place. It is impossible to articulate, in English, a dynamic world of interwoven grasses, trees, animals and water, fluctuating through time, as a single and eternal entity. It is also impossible to allow people into this process; we have no word for the progression and collaboration of the earth and all life, including humanity, evolving as one. English equips us with neat boxes of 'things' that engage in unidirectional, hierarchical interactions. This is not sufficient; we need a language of complexity and mutual dependence instead, one that emphasises connection and process instead of division and structure. One that is verb based instead of noun based. We need to craft a new language. One that has base words for processes; like wooding, grassing, nativing and exoticing. Words that can be built upon and refined over time to equip us with a vast dictionary of words to describe the flush of young birch after a woodland storm, the creep of willow carr or the stealthy march of blackthorn scrub as different types of wooding. Further words might stack on top of these to describe the passage of wind through different densities of canopy or the different textures of soil and leaf mould as they change with the trees and years.

We need a language, above all, that is capable of conceptualising and describing the world as a whole, all of the interactions between life and the processes that emerge. If we had a language that drew out the cyclical, ever changing nature of the world and described the importance of relationship and cooperation between all lives we might find it easier to create a culture that carries within it the knowing of the embeddedness of everything. If we understood and felt at home within a cyclical pattern of the world we might find it easier to trust in nature's abundance and loosen how tightly we cling to the notion of control and enforced permanence. Ultimately we might find it easier to both live, and die.

The Final Divide

With a new language it is possible to describe the whole within which human lives meld with the wild. Caroline Grindrod has coined the term Wilderculture to describe this whole and Rebecca

Hoskings, agriwilding. I use the term Holistic Restoration. Whatever we call it we are starting down a path of integration but this is a long road that stretches beyond the physical merging of people and wild. We can engineer the disturbance within the stability and create a language that ensures that these processes are seen as aspects of a single, whole landscape that mutually sustains itself but still there is a binary encapsulated within it. There is a divide between the areas where we exert our will and the wild matrix where nature's will holds sway. Human will is a powerful thing that encourages us to press against nature's feedback cycles, to push despite the warning signs at natural checks and balances, to act quickly before we feel the heavy pull of nature's hand correcting our course. We must be aware of this, understand just how fast we are, just how far our culture can push before we feel nature's resistance. We must be aware of our will as a process that extends out into the world and meets and merges with the larger processes of Gaia and the larger will of wild lands. At this boundary we need to shift our focus from competition and dominance to communication and respect; to work with life and direct our will towards sensing Gaia's feedback loops rather than trying to outsmart or break free of them. In order to fully heal our world we must reintegrate our will with that of nature so that they are complimentary forces, equally respected, once more.

Will is a fairly abstract concept but, just like language, has a tremendous impact on the real world around us. We currently have a very strong sense of our right to everything; our ability to do as we please and 'follow our dreams'. We are entering into relationship, however, and, just as in other relationships, the free will of the individual is compromised in exchange for the solidarity that comes from partnership. We must value the 'wild', not as a designation or a servant but as a process and a relationship. So far people have accepted letting wildlife into productive systems but we have not yet fully embraced the idea of one day relinquishing control of them. Everything changes and systems evolve, land cannot be bound up for ever more as fields or orchards governed by man. Rivers will rise, beavers will flood forests, people will shift and whole landscapes will fluctuate. We can participate in this dance but we cannot control it indefinitely. We must accept that we can co-sculpt fields of edible and wild species, forests of oaks and chestnuts, scrub of apples and hawthorn, self willed wood pastures of large herbivores but ultimately the dance is a synthesis of life and cannot be dictated by one species alone. We will have to give and take in equal measure. If we want to utilize the rich, fertile pastures built up by beavers we will have to allow other areas to be lost below the water leaving our old soils to be recharged by their careful hoarding of water and fertility. We will have to move on sometimes instead of automatically undoing natures work instead. More and less intensive human engineering may one day migrate across a landscape again on a long term cycle being cooperatively guided by humanity and nature over the centuries in respectful partnership.

This obviously doesn't work with current ideas around land ownership and is even less compatible with farmers whose lives and livelihoods are tied up in the value of the lands they work and produce they sell. I have no idea if people could be persuaded to pay for foods, contribute in taxes to subsidy schemes, or relinquish lands to level the playing field between farmers in more or less favourable areas of the landscape or sections of the dance. We must ensure the diversity of life and long term productivity of our landscapes across our island, however, and I believe that there is value in raising questions if only to stimulate thought and open up Overton's window.

To forge landscapes drawn from a united will, to act in partnership and acknowledge the mutual support, dependence and constraints that come with that relationship requires humility. We must see ourselves as more than individuals, as more than human, as a small part of a large community of life, as a member of no species and all species at once. And then surely what must emerge is the realisation that we are not separate, that our will is not separate, that our body is not separate but is all one. We are not discrete, we never were nor would we exist if we had been. We, and everything else, is one and the same. All life is one, and dances together or not at all.

Chapter 7

Emergence

It is our position as a global hyperkeystone, engineering species combined with how many of us there are that has brought such an array of crises to our planet, but those very same traits can move us towards restoration once more. The numbers of people alive in the world today, and destined to come into the world over the coming decades, are a tremendous asset if viewed as one. At our present population level we have managed to link vast numbers of people up around the globe and amazing things have emerged from this collaboration; the internet, machine learning, the ability to break the world down into its tiny subatomic units and at the same time gaze far out into space to witness the birth of galaxies. We are a fantastically inventive, curious and adaptable species. What we need at the moment is a colossal rethink of everything and we quite possibly have enough minds connected together around the world to achieve this. We also might very well have enough hands to make the skills and labour hungry regenerative farming systems discussed earlier, work. For the first time in history we might have a sufficient number of people on the planet to allow enough of us to relocate back into our landscapes; shepherding keystones, facilitating wooding and creating disturbance within stability, whilst retaining hubs of more intense human interaction. Dynamic, reimagined cities could facilitate the emergence of further experimentation and creativity to maintain a vibrant culture running on energy captured from the wind and sun and designed as a food web itself; every waste becoming the resource of another process. Cities themselves are just another ecosystem that we have engineered and could boost diversity, with their array of cliffs, mountain tops and deep ravines, if designed to do so. We need to reimage everything.

A supremely powerful, fast acting, flexible and numerous species could change the world very rapidly, and for the better, and climate change gives us a very real reason to focus our minds and do so. It is not, as Al Gore's book title says an 'Inconvenient Truth', it is a devastating one, but one that has the potential of being our salvation too. A consequence of our current relationship with the earth that is so big that it cannot be ignored. We probably need such a threat in order to re-evaluate to the depth needed and probe through the years of silently shifting baselines and imperceptibly changing water to come to a point where we are willing, and able, to unite around the world and imagine a different future.

One of Bill Mollison's original design principles of Permaculture was 'the problem is the solution'. But it needs a shift in perception to see it.

One thing that I repeatedly realise is that I know next to nothing about how our world works or who we are. After centuries of thought that has sought to divide up and break down we know virtually nothing about relationship or connectivity. After decades of seeing ourselves as individuals interacting with other individuals we know a lot about the individual but next to nothing about the interaction. I feel that the 'truth', if such a thing even exists, is always slightly beyond my reach, slightly too much, too big for me to comprehend. That it exists around and between everything, that it is an entirety whose fascinating complexity can be glimpsed in the network of mycorrhiza below a woodland floor or the dance of microbes between soils, guts and immune systems. I will never understand it and perhaps it is beyond knowing. I feel, however, that to move in the direction of incorporation and synthesis is to move in the direction of restoration. And that moving in a direction is the best that we can hope for, combined with a little faith in the unknowable complexity of life.

People seem to feel comfortable with linear systems, 1 2 3 or 2 4 6, logical patterns that are predictable and defined. They obey laws that we understand. But we seem to be a little hazy when it comes to non-linear systems. Even by naming them 'non-linear' there is perhaps the first warning sign that we have not truly grasped their magnitude, we have just grouped them as things that are not the sort of systems we usually talk about. They are not, however, just non-linear, they are totally different. They are unpredictable and uncontrollable. Life is a non-linear system and when we reintegrate aspects of the world we open the door to emergence, the massive unpredictability that springs from complexity. At every stage of this book we have encountered emergence. When grassing and wooding are united a massive level of productivity is unleashed. When diversity builds in ecosystems linking more and more species together; productivity, stability and resilience emerge. When we reintegrate wolves with deer we get dynamism that takes life and land off in totally unexpected ways. And if people are woven into the fabric of life once more we should not expect a simple addition of some wild and some farmed land, we should expect the emergence of something momentous. Emergence is a fundamental pattern of life and yet one that we are endlessly surprised by. There is absolutely no knowing where a path of reintegration, restoration and emergence might lead in a non-linear world but, in the face of what now threatens us, I find that reassuring. There is a real potential for all of the people around the world, interacting in complex and interdependent ways with the rest of life, to allow a world of restoration to emerge.

Humans exposed the earth to nuclear power; we let it leak out into our biosphere and then we ran before the threat of what we had unleashed, but in our wake, nature swelled. Now, 33 years on, the Pripyat River basin is recovering from the Chernobyl disaster. The Exclusion Zone hums with life as elk, deer, racoon dogs, foxes, boar, bison, horses and wolves roam the landscape once more. Even bears have ambled back.

We unleashed the most devastating side of ourselves and nature adapted. Imagine what could be accomplished if we unleashed the best of ourselves. If two sandhopper species living side by side on a wave tossed shore can break down more seaweed than either can on its own and the return of one keystone species to a mountain chain can alter the path of a river, imagine what could happen if the world's most powerful hyperkeystone species, its global engineer acknowledged itself as a part of the super organism that is Gaia. Imagine what might spring from such a collaboration around the world and what might transpire on our own small island. If farmers, rewilders and everybody else acknowledged that we are all just life bound together with a miraculous island on the edge of Europe imagine what that island might become. Imagine what that world might become. Imagine an Emergent World.

Conclusion

When I started writing this book, I did not know that it would be, in essence, a story of segregation and reintegration but I have realised that that is what it is. Our history, up until this point, has been characterised by segregation. We split our landscapes off from their vast sculptors when we consigned the megafauna to extinction. We segregated people from the lands that they used to engineer when we enclosed them to enhance productivity. We gradually chipped away at life, removing species after species from our productive system to be left with only the fastest few. We sundered plants from soil, grassland from woodland, animal from plant so that all might be grown in the isolation of a barn, a field, a plantation. We have systematically unpicked the relationships that held our world together and we are now experiencing the instability that has arisen from the disjointed collection of those that remain.

It is not the first time that people have destabilised our world. When humans moved out of Africa millions of years ago and spread, in waves of various species of *Homo*, across Eurasia they most likely impacted the ecosystems that they found. When *Homo sapiens* moved out of Africa about 70,000 to 100,000 years ago we took the game up a notch. We are an incredibly interactive and powerful species; more powerful than prairie dogs or bees, more powerful even than elephants when it comes to sculpting landscapes, and when we arrive, we arrive with a bang. As people spread, species vanished. Across Eurasia we were treading in the footsteps of the previous species of *Homo* and they vanished about us along with elephants, mammoths and rhinos. When we moved beyond the fringes of *Homo* territory, into Australia and then the Americas, where creatures had never experienced any member of our genus before, food webs crumbled. Our arrival fundamentally reshuffled the ecosystems of the planet and new collections of species and landscapes emerged.

The history of the colonisation of the Americas and the brutal displacement of Native Americans by Europeans is well known and it is widely accepted that prior to European settlement the Native Americas lived in harmony with their environment. Sitting alongside this understanding is the knowledge gained from various archaeological dig sites across both North and South America that suggests that the New World lost a colossal quantity of species including mastodons, mammoths, a six foot long capybara and giant beavers when the people who would go on to become Native Americans first arrived. Unsurprisingly there is a tension between the two accounts; how could the people who caused so many species to go extinct about 14,000 years ago be regarded as people who lived in respectful balance with pre European conquest ecosystems? There is ecological room, however, for the two stories to coexist peacefully with one another.

The first people to make their way into the Americas could very well have sent ripples out through the food web that destabilised relationships and resulted in the extinctions of many species. Ten thousand years later, however, it is also entirely possible that those ripples would have died down. That when given enough time the remaining species, people included, evolved and adapted. Food webs would have moved towards new stable states. The accounts of some of the first Europeans exploring North America described large parkland expanses of mature trees and open grass grazed by deer, elk and bison. Swathes of blackberries and stands of oaks, hickories and chestnuts laced with turkeys and bears. They described an ecosystem in which a powerful keystone species, not an elephant but a person, had entered into a balance with landscapes and other species. Conjuring trees from the grassland in one place and removing them in another, people kept landscapes on the move.

The two stories, of mass extinction and balanced relationship, are very possibly the same story told from different time periods. When there is a change, a disturbance, when a species is dropped in or pulled out of an ecosystem too quickly food webs tremble from the shock. Over time, however, the new presence or absence is adapted to, food webs shift and evolve and a new stability emerges.

Britain was a steady land of scrub, elephants, rhinos and Neanderthals that suffered a blow as the elephants, rhino and Neanderthals slipped into extinction, rattled from their landscapes by *Homo sapiens*. But the scrub persisted, as did most of our other wildlife, shifting into landscapes left in the shadow of humans instead of elephants. Slowly the munching mouths of boar and aurochs were swapped for the domesticated foraging of pigs and cattle but still our overall landscape persisted as woods, grasslands, wood pastures and wetlands. Thousands of years of steady eating moved Britain from the end of the last ice age to the loose farms of a few hundred years ago. Food webs morphed gradually but feedback loops still bound life together and knitted life into a mottled mosaic of scrub, wood and water which functioned in a similar way.

Then we evolved for the second time. Not a biologically distinct species but a culturally distinct one, born in the smoky cities of England, we dispersed again as industrial *Homo sapiens*. Over the last 250 years or so almost every aspect of human life has changed across vast areas of the world incredibly rapidly. We have changed who we are ecologically and that has had a huge effect on the food globe but history shows us that we can move from this extinction causing whirlwind into a foundational keystone, underpinning diversity. We know this because we have already done it. All over the world. In all sorts of places from the humid tropics to the great grasslands. We can, and have been, keystone species and ecosystem engineers.

The main strand that runs through the history of our species' time on earth is one of repeatedly segregating and destabilising relationships before being swept up and stabilised by them once more. Unfortunately we don't have the thousands of years that our ancestors benefitted from after their initial introduction to refind balance. We also don't have to wait to passively drift, guided by nature's feedback loops and human observation, towards stability as they did. We now have an enhanced level of understanding about the world, better technology and more people. We have a global level of insight and tremendous power. We can design restoration into our lives and we can set up and monitor systems to ensure that we are headed in the right direction. We can restore the relationships between soil and water and woodlands and grasslands, or between landscapes with their great keystone sculptors including ourselves. We can recreate complexity and assume the emergence of *something*, but that something can never be known until it appears. An emergent property, by definition, is more than the sum of its parts and currently we only have the parts. We can never know what, if anything, will emerge which is what makes restoration ultimately a work of hope and faith. Hope that it is not too late to repair the damage that we have done and faith in the global web of life to move towards a state that fosters further life, of which we are but one tiny thread.

Author Biography

Miriam Kate McDonald has over 15 years of experience in ecology, conservation and agriculture working in some of Britain's most beautiful landscapes. She created and co-directs Holistic Restoration with Robert Owen from the University of Exeter, exploring and recreating humanity's ecological connections with the wild through research, teaching and consultancy.

www.holisticrestoration.co.uk

References and Further Reading

Chapter 1. The Changing Face of Britain

Martin, P S. 2007. *Twilight of the mammoths: ice age extinctions and the rewilding of America*. University of California Press

Flannery, S. 2018. *Europe the first 100 million years*. Allen Lane

Nisbet, R E. 1991. *Living earth: a short history of life and its home*. Springer

Lenton, T and WATSON, A. 2013. *Revolutions that made the earth*. OUP Oxford

Nogues-Bravo, D. Rodrigues, J. Hortal, J. Batra, P and Araujo, M. 2008. *Climate change, humans and the extinction of the woolly mammoth*. PLoS Biol 6(4)

Crane, N. 2017. *The making of the British landscape*. Weidenfeld & Nicholson

Rackham, O. 2001. *Trees and woodland in the British landscape*. W&N

Rackham, O. 2000. *The history of the countryside*. W&N

Whitefield, P. 2009. *The living landscape: how to read and understand it*. Permanent publications

Rodwell, J. 2017. *The UK national vegetation classification system*. Phytocoenologia 48(2)

Rodwell, J S. 2008. *British plant communities volume 1*. Cambridge University Press.

Rodwell, J S. 2008. *British plant communities volume 2*. Cambridge University Press.

Rodwell, J S. 2008. *British plant communities volume 3*. Cambridge University Press.

Rodwell, J S. 2008. *British plant communities volume 4*. Cambridge University Press.

Mazoyer, M. 2006. *A history of world agriculture: from the Neolithic age to the current crisis*. Earthscan

Vasey, D E. 1990. *An ecological history of agriculture, 10,000 BC; AD 10,000*. Iowa State University Press

Higman, B W. 2011. *How food made history*. Wiley-Blackwell

Diamond, J. 1998. *Guns, germs and steel: a short history of everybody for the last 13,000 years*. Vintage

Diamond, J. 2011. *Collapse: how societies choose to fail or survive*. Penguin

Harari, Y N. 2015. *Sapiens: a brief history of humankind*. Vintage

Tudge, C. 1997. *The day before yesterday: five million years of human history*. Pimlico

COHEN, M N and ARMELAGOS, G J. 2013 (2nd ed). *Paleopathology at the origins of agriculture*. University Press of Florida.

Rindos, D. 1987. *The origins of agriculture: an evolutionary perspective*. Academic Press

Richerson, P J and Boyd, R. 2006. *Not by genes alone: how culture transformed human evolution*. University of Chicago Press

Lent, J. 2017. *The patterning instinct: a cultural history of humanity's search for meaning*. Prometheus Books

Descartes, R. 1644. *Principles of philosophy*. SMK Books 2009.

Shrubsole, G. 2019. *Who owns England? How we lost our green and pleasant land and how to take it back*. William Collins

Ryder, M L. 1964. *The history of sheep breeds in Britain*. The Agricultural History Review 12(1): 1-12

Humphries, A B. 2015. *Hill sheep husbandry in England: Adaptive to change in diverse ecosystems*. Cumbria: Foundation for Common Land

Fairlie, S. 2009. *A short history of enclosure in Britain*. The Land Magazine, Summer 2009

Hardin, G. 1968. *The tragedy of the commons*. Science :1243-1248.

Lloyd, W F. 1833. *Two lectures in the checks to population*. Oxford University Press

Hoffmann, R. 2014. *An environmental history of medieval Europe*. Cambridge University Press

Simmons, I G. 2001. *An environmental history of Great Britain: from 10,000 years ago to the present.* Edinburgh University Press

Overton, M. 1996. *Agricultural revolution in England: the transformation of the agrarian economy 1500- 1850.* Cambridge University Press

Yallop, A R. Thomas, G. Thacker, J. Brewer, T and Sannier, C. 2005. *A history of burning as a management tool in the English uplands.* English Nature Research Report 667

Yallop, A R. Thomas, G. Thacker, J. Stephens, M. Clutterbuck, B. Brewer, T and Sannier, C. 2006. *The extent and intensity of management burning in the English uplands.* Journal of Applied Ecology 43(6): 1138-1148

Carson, R. 2003. *Silent spring.* Houghton Mifflin 40th anniversary edition

Van Den Bosch, R. 1989. *The pesticide conspiracy.* University of California Press

Mcneill, J R. 2001. *Something new under the sun: An environmental history of the twentieth century world.* W. W. Norton and Company

ed Brassley, P. Segers, Y. Van Molle, L. 2012. *War, agriculture and food: rural Europe from the 1930s to the 1950s.* Routledge

Environmental challenges in farm management (ECIFM). 2012. *Agriculture in post war Britain.* University of Reading. http://www.ecifm.rdg.ac.uk/

Rogosa, E. 2016. *Restoring heritage grains: the culture, diversity and resilience of landrace wheat.* Chelsea Green Publishing Company

Chapter 2. Ripples of Change

Howard, A. 2010. *An agricultural testament.* Benediction Classics

Lockeretz, W. 2011. *Organic farming: an international history.* CABI Publishing

Balfour, E B. 1951. *The living soil; evidence of the importance to human health of soil vitality*. Faber and Faber

Northbourne, W. 2004. *Look to the land.* Angelico Press/Sophia Perennis

Wordsworth, W. 2010. *The complete poetical works of William Wordsworth.* Nabu Press

Sands, T. 2012. *Wildlife in trust: a hundred years of nature conservation.* Elliott and Thompson Limited

Evans, D. 1997. *A history of nature conservation in Britain, second edition.* Routledge

Rees, R. 1982. *Constable, Turner, and Views of Nature in the Nineteenth Century.* Geographical Review 72(3): 253-269

Jan Oosthoek, K. 2015. *Romanticism and nature.* Environmental History Resources

Perkins Marsh, G. 2003. *Man and nature: or, physical geography as modified by human action.* University of Washington Press; reprint edition.

Henderson, N. 1992. *Wilderness and the nature conservation ideal: Britain, Canada, and the United States contrasted.* Ambio 21(6): 394-399

Thoreau, H. 2011. *The journal. 1837-1861.* NYRB Classics.

Zahniser, H. 1964. *The wilderness act.* Pub.L. 88-577

Wulf, A. 2016. *The invention of nature: the adventures of Alexander von Humboldt, the lost hero of science.* John Murray

Hagen, J B. 1992. *An entangled bank: origins of ecosystem ecology.* Rutgers University Press

Lovelock, J. 2000. *Gaia: a new look at life on earth.* OUP Oxford 3rd edition

Lovelock, J. 2007. *The revenge of Gaia: why the earth is fighting back and how we can still save humanity.* Penguin

Meadows, D H. 1972. *The limits to growth.* Signet

Caradonna, J L. 2016. *Sustainability: a history.* Oxford University Press

Luisi, P L and Capra, F. 2016. *The systems view of life.* Cambridge University Press

Briggs, J. 1992. *Fractals: The patterns of chaos*. Pocket Books

Marsh, S J. 2008. *Biodiversity' and ecosystem processes in the strandline: The role of species identity, diversity, interactions and body size*. PhD thesis at the University of Plymouth

Barrow-Green, J. 1997. *Poincare and the three body problem*. American Mathematical Society.

Werner, R.F. and Farrelly, T. 2019. *Uncertainty from Heisenberg to today*. arXiv:1904

Chapter 3. Natures Patterns

Stanley, S M. 2014. *Earth system history*. W. H. Freeman and Co 4th edition

Kump, L R. Kasting, J F and Crane, R G. 2003. *The earth system: an introduction to earth systems science*. Pearson 2nd edition

Wilsson, L and Bulman, J. 1969. *My beaver colony*. Souvenir P

Townsend, C R. Begon, M and Harper J L. 2008. *Essentials of ecology*. Wiley and Blackwell 3rd edition

Hairston, N. Smith, F and Slobodkin, L. 1960. *Community structure, population control and competition*. The American Naturalist 94(879)

Estes, J.A. Dayton, P.K. and Kareiva, P. 2016. *A keystone ecologist: Robert Treat Paine*. Ecology 97(11)

Townsend, C R. Begon, M and Harper J L. 2006. *Ecology: from individuals to ecosystems*. John Wiley and Sons 4th edition

Spedding, C R W. 1976. *Grassland ecology*. Oxford University Press

Gibson, D J. 2009. *Grasses and grassland ecology*. Oxford University Press

Blakesley, D and Buckley, P. 2016. *Grassland restoration and management*. Pelagic Publishing

Putman, R J. 2013. *Grazing in temperate ecosystems: large herbivores and the ecology of the New Forest*. Springer 2nd edition

Thomas, P A and Packham, J R. 2007. *Ecology of woodlands*

and forests: description, dynamics and diversity. Cambridge University Press

Vera, F. 2000. *Grazing ecology and forest history.* CABI Publishing

Lafortezza, R. Chen, J. Sanesi, G and Crow, T R. 2008. *Patterns and processes on forest landscapes: multiple use and sustainable management.* Springer

Van Der Valk, A G. 2012. *The biology of freshwater wetlands (biology of habitats series).* Oxford University Press 2nd edition

Bronmark, C and Hansson, L A. 2005. *The biology of lakes and ponds (biology of habitats series).* Oxford University Press 2nd edition

Giller, P S. 1999. *The biology of streams and rivers (biology of habitats series).* Oxford University Press

Dobson, M and Frid, C. 2008. *Ecology of aquatic systems.* OUP Oxford 2nd edition

Fornara, D and Tilman, D. 2008. *Plant functional composition influences rates of soil carbon and nitrogen accumulation.* Journal of Ecology 96(2)

Semlitsch, R D. O'donnell, K M and Thompson, F R. 2014. *Abundance, biomass production, nutrient content, and the possible role of terrestrial salamanders in Missouri Ozark forest ecosystems.* Can. J. Zool. 92: 997–100

Monbiot, G. 2019. *Averting climate breakdown by restoring ecosystems. A call to action.* Natural Climate Solutions Full Rationale. Available at https://www.naturalclimate.solutions/full-rationale

Alonso, I. Weston, K. Gregg, R and Morecroft, M. 2012. *Carbon storage by habitat: a review of the evidence of the impacts of management decisions and condition of carbon stores and sources.* Natural England Research Report NERR043

Rewilding and climate breakdown: how restoring nature can help decarbonise the UK. Rewilding Britain. 2019.

Griscom, B. et al. 2017. *Natural climate solutions.* PNAS 114(44): 11645-11650

Hodkinson, I D and Coulson, S.J. 2004. *Are high arctic terrestrial food chains really that simple? The bear island food web revisited.* Oikos 106(2)

Lake, S. Liley, D. Still, R and Swash, A. 2014. *Britain's habitats: a guide to the wildlife and habitats of Britain and Ireland.* Princeton University Press

Darwin, C. 1998. *On the origin of species (classics of world literature).* Wordsworth Editions

Tudge, C. 2013. *Why genes are not selfish and people are nice: a challenge to the dangerous ideas that dominate our lives.* Floris Books

Cazzola Gatti, R. 2016. *A conceptual model of a new hypothesis on the evolution of biodiversity.* Biologia 71(3): 343-351

Estes, J. 2016. *Serendipity: An ecologist's quest to understand nature.* University of California Press

Carroll, S. 2016. *The Serengeti rules: The quest to discover how life works and why it matters.* Princeton University Press

Goldfarb, B. 2018. *Eager: the surprising, secret life of beavers and why they matter.* Chelsea Green Publishing Company

Arthur, R. 2019. *Seed banks, entropy and Gaia.* https://arxiv.org/pdf/1907.12654.pdf

Simberloff, D S and Wilson, E O. 1969. *Experimental zoogeography of islands: the colonisation of empty islands.* Ecology 50(2)

Chapter 4. Rewilding Conservation

Macarthur, R H and Wilson, E O. 1967. *The theory of island biogeography.* Princeton University Press

Macdonald, G. 2002. *Biogeography: introduction to space, time and life.* John Wiley and Sons

Palmer, M A. Zedler, J B and Falk, D A. 2017. *Foundations of restoration ecology, second edition.* Island Press

Muir, J. 1911. *My first summer in the Sierra.* Houghton Mifflin.

Friederici, P. 2006. *Nature's restoration: people and places on the front lines of conservation.* Island Press

Jordan, W R. 2003. *The sunflower forest.* University of California Press

Sandom, C. Clouttick, D. Manwill, M and Bull, J W. 2016. *Rewilding knowledge hub: bibliography; version 1.* Rewilding Britain

Macdonald, B. 2019. *Rebirding: rewilding Britain and its birds.* Pelagic Publishing

Fraser, C. 2010. *Rewilding the world: dispatchers from the conservation revolution.* Picador USA

Tree, I. 2019. *Wilding: the return of nature to a British farm.* Picador.

Barnett, R. 2019. *The missing lynx, the past and future of Britain's lost mammals.* Bloomsbury Wildlife

Wiersma, Y and Sandlos, J. 2011. *Once there were so many: animals as ecological baselines.* Environmental History 16(3): 400-407

Minteer, B. 2019. *The fall of the wild: Extinction, de-extinction and the ethics of conservation.* Columbia University Press

Soule, M and Noss, R. 1998. *Complementary goals for continental conservation.* Wild Earth 8(3) 18-28

Ripple, W J and Beschta, R.L. 2007. *Restoring Yellowstone's aspen with wolves.* Biological Conservation 138(3-4)

Elton, C. 2000. *The ecology of invasions by animals and plants.* University of Chicago Press

Lampert, A. Hastings, A. Grosholz, E and Jardine, S. 2014. *Optimal approaches for balancing invasive species eradication and endangered species management.* Science 344 (6187)

Whelan, C. 2013. *Spotlight on sustainability: The importance of ecotourism.* World Wildlife Fund

Balmford, A. Green, M H. Anderson, M. Beresford, J. Huang, C. Naidoo, R. Walpole, M and Manica, A. 2015. *Walk on the wild side: estimating the global magnitude of visits to protected areas.* PLoS Biol 13(2)

Monbiot, G. 2014. *Feral: rewilding the land, sea and human life.* Penguin

Henderson, D. *American Wilderness Philosophy*. The Internet Encyclopaedia of Philosophy, ISSN 2161-0002, https://www.iep.utm.edu/, 15/12/2019.

Nash, R F. 2014. *Wilderness and the American mind 5th ed.* New Haven

Ward, K. 2019. *For wilderness or wildness? Decolonising rewilding.* Plymouth University

Whatmore, S and Thorne, L. 1998. *Wild(er)ness: Reconfiguring the geographies of wildlife.* Transactions of the Institute of British Geographers 23(4)

Worm, B and Paine, R T. 2016. *Humans as hyperkeystone species.* Trends in Ecology and Evolution 31(8): 600-607

Root-Bernstein, M and Ladle, R J. 2019. *Ecology of a widespread large omnivore,* Homo sapiens, *and its impacts on ecosystem processes.* Ecology and Evolution 9(1625)

Dunne, Ja. Maschner, H. Betts, Mw. Hunty, N and Russell, R. 2016. *The roles and impacts of human hunter-gatherers in North Pacific marine food webs.* Scientific Reports 6

Sutherland, W J. 1995. *Managing habitats for conservation.* Cambridge University Press

Wright, J P. Jones, C G and Flecker, A S. 2002. *An ecosystem engineer, the beaver, increases species richness at the landscape level.* Oecologia 132: 96-101

Haemig, P D. 2012. *Ecosystem engineers: organisms that create, modify and maintain habitats.* Ecology

Boivin, N L. Zeder, M A. Fuller, D Q, Crowther, A. Larson, G. Erlandson, J M. Denham, T. and Petraglia, M D. 2016. *Ecological consequences of human niche construction: Examining long-term anthropogenic shaping of global species distributions.* PNAS 113(23): 6388-6396

Bocherens, H. 2018. *The rise of the Anthroposphere since 50,000 years: An ecological replacement of megaherbivores by humans in terrestrial ecosystems.* Front. Ecol. Evol.

Mann, C. 2005. *1491: new revelations of the America before Columbus.*

Knopf Publishing Group

Anderson, M.K. 2013. *Tending the wild: Native American knowledge and the management of California's natural resources*. University of California Press

Gammage, B. 2012. *Biggest estate on earth: How aborigines made Australia*. Allen and Unwin

The Integrative HMP Research Network Consortium. Proctor, L.M. et al. 2019. *The integrative human microbiome project*. Nature 569

Roth, J. Leroith, D et al. 1982. *The evolutionary origins of hormones, neurotransmitters and other extracellular chemical messengers - implications for mammalian biology*. The New England Journal of Medicine.

Montgomery D R. and Bikle, A. 2015. *The hidden half of nature: the microbial roots of life and health*. W. W. Norton and Company

Mayer, E. 2016. *The mind-gut connection: How the hidden conversation within our bodies impacts our mood, our choices and our overall health*. Harper Wave

Worth, R N and Burnard, R. 2006. *General karst environmental overview of the Cattedown area of Plymouth*. The Devon Karst Research Society

Plymouth Sound and Tamar Estuaries MPA. 2014. *Habitats and wildlife around the MPA*. www.plymouth-mpa.uk/home/about/plymouth-sound-tamar-estuaries/#1532352094051-274b9cf0-8042

Ware, S and Meadows, B. 2012. *Monitoring of Plymouth Sound and estuaries SAC 2011*. Cefas final report

Chapter 5. Regenerating Agriculture

Grain. 2014. *Hungry for land: small farmers feed the world with less than a quarter of all farmland*. https://www.grain.org/article/entries/4929-hungry-for-land-small-farmers-feed-the-world-with-less-than-a-quarter-of-all-farmland

Pearce, F. 2018. *Sparing vs. sharing: the great debate over how to protect nature.* Yale School of Forestry and Environmental Studies

Policy Foresight Programme. 2015. *Can Britain feed itself? Should Britain feed itself?* https://www.oxfordmartin.ox.ac.uk/downloads/reports/PFPShould_Britain_Feed_Itself.pdf

Fairlie, S. 2007. *Can Britain feed itself?* The Land 4 Winter 2007-8

Fairlie, S. 2010. *Meat: a benign extravagance.* Permanent publications

Bongaarts, J. 1994. *Can the growing human population feed itself?* Scientific American. Mar. 1994: 36-42.

Cohen, J E. 1995. *Population growth and earth's human carrying capacity.* Science 269: 341-346.

Evans, L T. 1998. *Feeding the ten billion. Plants and population growth.* Cambridge University Press.

FAO. 2019. *Crop prospects and food situation, quarterly global report for the food and agriculture organisation of the United Nations.* FAO

Office of National Statistics. *UK environmental accounts: 2019* available at https://www.ons.gov uk/economy/environmentalaccounts/bulletins/ukenvironmentalaccounts/2019

UK National Ecosystem Assessment 2011 available at uknea.unep-wcmc.org/

Altieri, M.A. Funes-Monzote, F.R. and Petersen, P. 2012. *Agroecologically efficient agricultural systems for smallholder farmers: contributions to food sovereignty.* Agron. Sustain. Dev. 32

Chettri, B. 2015. *Organic farming in Sikkim: Implications for livelihood diversification and community development.* Thesis submitted to the Department of Economics Sikkim University.

Herve-Gruyer, C. 2016. *Miraculous abundance: one quarter acre, two French farmers and enough food to feed the world.* Chelsea Green Publishing Company

Holzer, S. 2010. *Sepp Holzer's Permaculture.* Permanent

Publications

Fern, K. 2011. *Plants for a future: edible and useful plants for a healthier world*. Permanent Publications 2nd edition

Yeomans, P A. 2008. *Water for every farm: Yeomans keyline plan*. Createspace Independent Publishing Platform

King, F H. 2016. *Farmers of forty centuries*. CreateSpace Independent Publishing Platform

Jadrnicek, S. 2016. *The bio-integrated farm and home*. Chelsea Green Publishing Company

Southern, A and King, W. 2017. *The aquaponic farmer: a complete guide to building and operating a commercial aquaponic system*. New Society Publishers

Bernstein, S. 2013. *Aquaponic gardening: a step-by-step guide to raising vegetables and fish together*. Saraband

Falk, B. 2013. *The resilient farm and homestead: an innovative permaculture and whole system design approach*. Chelsea Green Publishing Company

Global Food Security. 2015. *Global food security workshop: Insects as animal feed.*

Nardi, J B. 2007. *Life in the soil: a guide for naturalists and gardeners*. University of Chicago Press

Lowenfels, J and Lewis, W. 2010. *Teaming with microbes*. Timber Press

Phillips, M. *Mycorrhizal planet: how symbiotic fungi work with roots to support plant health and build soil fertility*. Chelsea Green Publishing Company

Kourik, R. 2015. *Understanding roots: discover how to make your garden flourish*. Metamorphic Press

Bardgett, R. 2005. *The biology of soil: a community and ecosystem approach (biology of habitats)*. Oxford University Press

Edmondson, J L. Davies, Z G. Gaston, K J and Leake, J R. 2014. *Urban cultivation in allotments maintains soil qualities adversely affected by conventional agriculture*. Journal of Applied Ecology Volume 51(4)

Arsenault, C. 2014. *Only 60 years of farming left if soil degradation continues.* Scientific American

Krebs, C. 2019. *Soil ecologist challenges mainstream thinking on climate change.* Regeneration International

Moyer, J. 2011. *Organic no-till farming.* Acres USA

Mefferd, A. *The organic no-till farming revolution: high-production methods for small scale farmers.* New Society Publishers

Grime, J P. 2012. *The evolutionary strategies that shape ecosystems.* Wiley-Blackwell

The Land Institute. *Perennial crops: new hardware for agriculture.* Available at https://landinstitute.org/our-work/perennial-crops/

Smith, J R. 2013. *Tree crops a permanent agriculture.* Island Press

Greencuisine Trust. 2016. *Five crops that could be integrated beneficially into UK growing systems.* Available at http://www.greencuisinetrust.org/communities/3/004/013/524/783/files/4632016036.pdf

Crawford, M. 2016. *How to grow your own nuts: choosing, cultivating and harvesting nuts in your garden.* Green Books

Perkins, R. 2016. *Making small farms work.* Ridgedale Permaculture

Savory, A and Butterfeild, J. 2016. *Holistic management: a commonsense revolution to restore our environment.* Island Press

Geremia, C. Merkle, J A. Eacker, D R. Wallen, R L. White, P J. Hebblewhite, M. and Kauffman, M J. 2019. *Migrating bison engineer the green wave.* PNAS

Nordborg, M. 2016. *Holistic management; a critical review of Allan Savory's grazing method.* Centre for Organic Food & Farming & Chalmers

pangala, S R. 2014. *Methane emissions from wetland trees: controls and variability.* PhD thesis, The Open University.

Becker, W. Kreuter, U P. Atkinson, S F. and Teague, W R. 2017. *Whole-ranch unit analysis of multipaddock grazing on rangeland sustainability in North Central Texas.* Rangeland Ecology & Management 70(4)

Briske, D D. Sayre, N F. Huntsinger, L. Fernandez-Gimenez, M. Budd, B and Derner, J D. *Origin, persistence and resolution of the rotational grazing debate: integrating human dimensions into rangeland research.* Rangeland Ecology and Management 64(4): 325-334

Dittman, M. *The future of farming. Lowering methane emissions.* University of Reading

Hammond, K J. Humphries, D J. Westbury, D B. Thompson, A. Crompton, L A. Kirton, P. Green, C. and Reynolds, C K. 2014. *The inclusion of forage mixtures in the diet of growing dairy heifers: Impacts on digestion, energy utilisation, and methane emissions.* Agriculture, Ecosystems & Environment 197: 88-95

Ekarius, C. 1999. *Small-scale livestock farming.* Storey Books

Ussery, H. 2013. *The small-scale poultry flock: an all-natural approach to raising chickens and other fowl for home and market growers.* Chelsea Green Publishing Company

Salatin, J. 1996. *Pastured poultry profits: net $25,000 in 6 months on 20 acres.* Polyface, Incorporated

Salatin, J. 1996. *Salad bar beef.* Polyface, Incorporated

Crawford, M. 2010. *Creating a forest garden: working with nature to grow edible crops.* Green Books

Wohlleben, P. 2017. *The hidden life of trees: how they feel, how they communicate.* William Collins

Gordon, A M. Newman, S M and Coleman, B. 2018. *Temperate agroforestry systems.* CABI, 2nd edition

Steppler, H A. 1987. *Agroforestry: a decade of development.* Agribookstore

Rigueiro-Rodriguez, A. Mcadam, J. and Mosquera-Losada, M R. 2009. *Agroforesrty in Europe: current trends and future prospects.* Springer

Jacke, D. 2005. *Edible forest gardens: volumes 1 and 2.* Chelsea Green Publishing Company

Phillips, M. 2012. *The holistic orchard: tree fruits and berries the biological way.* Chelsea Green Publishing Company

Mollison, B. 1988. *Permaculture: a designer's manual.* Tagari

Whitefield, P. 2016. *The earth care manual: a permaculture handbook for Britain and other temperate climates.* Permanent Publications

Shepard, M. 2013. *Restoration agriculture.* Acres USA

Fukuoka, M. 2009. *The one straw revolution.* NYRB Classics

Fukuoka, M. 1985. *The natural way of farming: the theory and practise of green philosophy.* Bookventure

Pollan, M. 2011. *The omnivore's dilemma: the search for a perfect meal in a fast-food world.* Bloomsbury Paperbacks

Janssen, M A. 2001. *An immune system perspective on ecosystem management.* Conservation Ecology 5(1): 13

Gliessman, S R. 2014. *Agroecology: the ecology of sustainable food systems, third edition.* CRC Press

Wojtkowski, P. 2003. *Landscape agroecology.* CRC Press

Chapter 6. Integration

Gore, A. 2007. *An inconvenient truth: The crisis of global warming.* Perfection Learning

Appleton, J. 1996. *The experience of landscapes.* Wiley-Blackwell

Woodcock, B A. Bulock, J M. McCracken, M and Chapman, R E. 2016. *Spill-over of pest control and pollination services into arable crops.* Agriculture Ecosystems & Environment 231:15-23

Schneider, G. 2015. *Effects of adjacent habitats and landscape composition on biodiversity in semi-natural grasslands and biological pest control in oilseed rape fields.* PhD thesis at the University of Würzburg

Perfecto, I. Vandermeer, J. and Wright, A. 2019. *Nature's matrix: linking agriculture, biodiversity conservation and food sovereignty.* Routledge

Abraham, A. 2019. *Queer intentions: A (personal) journey through LGBTQ+ culture.* Picador

Barker, M.J. and Iantaffi, A. 2020. *Life isn't binary: On being both, beyond and in-between.* Jessica Kingsley Publishers

Everett, D. 2013. *Language: The cultural tool*. Profile Books

Pinker, S. 1995. *The language instinct: How the mind creates language*. Penguin.

**EARTH
BOOKS**

ENVIRONMENT

Earth Books are practical, scientific and philosophical publications about our relationship with the environment. Earth Books explore sustainable ways of living; including green parenting, gardening, cooking and natural building. They also look at ecology, conservation and aspects of environmental science, including green energy. An understanding of the interdependence of all living things is central to Earth Books, and therefore consideration of our relationship with other animals is important. Animal welfare is explored. The purpose of Earth Books is to deepen our understanding of the environment and our role within it. The books featured under this imprint will both present thought-provoking questions and offer practical solutions. If you have enjoyed this book, why not tell other readers by posting a review on your preferred book site.

Safe Planet
Renewable Energy Plus Workers' Power
John Cowsill
Safe Planet lays out a roadmap of renewable energy sources and meteorological data to direct us towards a safe planet.
Paperback: 978-1-78099-682-0 ebook: 978-1-78099-683-7

Approaching Chaos
Could an Ancient Archetype Save 21st Century Civilization?
Lucy Wyatt
Civilisation can survive by learning from the social, spiritual and technological secrets of ancient civilisations such as Egypt.
Paperback: 978-1-84694-255-6

Gardening with the Moon & Stars
Elen Sentier
Organics with Ooomph! Bringing biodynamics to the ordinary gardener.
Paperback: 978-1-78279-984-9 ebook: 978-1-78279-985-6

GreenSpirit
Path to a New Consciousness
Marian Van Eyk McCain
A collection of essays on 21st Century green spirituality and its key role in creating a peaceful and sustainable world.
Paperback: 978-1-84694-290-7 ebook: 978-1-78099-186-3

The Protein Myth
Significantly Reducing the Risk of Cancer, Heart Disease, Stroke, and Diabetes While Saving the Animals and the Planet
David Gerow Irving
The Protein Myth powerfully illustrates how the way to vibrant health and a peaceful world is to stop exploiting animals.
Paperback: 978-1-84694-673-8 ebook: 978-1-78099-073-6

This Is Hope
Green Vegans and the New Human Ecology How We Find Our
Way to a Humane and Environmentally Sane Future
Will Anderson
This Is Hope compares the outcomes of two human ecologies;
one is tragic, the other full of promise...
Paperback: 978-1-78099-890-9

Readers of ebooks can buy or view any of these bestsellers by
clicking on the live link in the title. Most titles are published
in paperback and as an ebook. Paperbacks are available in
traditional bookshops. Both print and ebook formats are
available online.

Find more titles and sign up to our readers' newsletter at
http://www.johnhuntpublishing.com/non-fiction
Follow us on Facebook at https://www.facebook.com/
JHPNonFiction